# Climate Smart Agriculture

**Scrivener Publishing**
100 Cummings Center, Suite 541J
Beverly, MA 01915-6106

*Publishers at Scrivener*
Martin Scrivener (martin@scrivenerpublishing.com)
Phillip Carmical (pcarmical@scrivenerpublishing.com)

# Climate Smart Agriculture

Edited by

**Anitha Velu**

**Prasanth Aruchamy**

**Raghu Ramamoorthy**

**Rajesh Kumar Dhanaraj**

and

**Seifedine Kadry**

Scrivener Publishing

WILEY

This edition first published 2026 by John Wiley & Sons, Inc., 111 River Street, Hoboken, NJ 07030, USA and Scrivener Publishing LLC, 100 Cummings Center, Suite 541J, Beverly, MA 01915, USA

For more information about Scrivener publications please visit www.scrivenerpublishing.com.



**Wiley Global Headquarters**
111 River Street, Hoboken, NJ 07030, USA

For details of our global editorial offices, customer services, and more information about Wiley products visit us at www.wiley.com.

The manufacturer's authorized representative according to the EU General Product Safety Regulation is Wiley-VCH GmbH, Boschstr. 12, 69469 Weinheim, Germany, e-mail: Product_Safety@wiley.com.



***Library of Congress Cataloging-in-Publication Data***

ISBN 9781394336333

Cover image: Generated with AI using Adobe Firefly
Cover design by Russell Richardson

Set in size of 11pt and Minion Pro by Manila Typesetting Company, Makati, Philippines

Printed in the USA

10 9 8 7 6 5 4 3 2 1

# Contents

# Preface

This book *Climate Smart Agriculture* offers a thorough investigation of how sophisticated computational intelligence, may turn agriculture into a data-driven, climate-resilient, and sustainable industry. This book connects fundamental theory with real-world application which gives readers a solid conceptual framework by looking at deep reinforcement learning from the viewpoints of optimal control and artificial intelligence. The book builds on this basis by examining the fundamentals and its uptake, effects, and wider ramifications of Climate Smart Agriculture for sustainable development. The book shows how adaptive decision-making may maximise productivity while preserving natural resources by fusing state-of-the-art algorithms.

The multidisciplinary integration of this book is one of its unique contributions. Agroforestry conservation, soil microbiome management, AR/VR-assisted modelling, and controlled environment agriculture are a few examples of how immersive technologies and intelligent agents can improve environmental stewardship. High-tech greenhouses, vertical farms, and scalable grain farming systems are highlighted in particular, along with issues with data privacy, cybersecurity, and system scalability. The practical usefulness is strengthened by the inclusion of thorough case studies. Each case study demonstrates practical uses of reinforcement learning, ranging from yield prediction using reliable data mapping approaches to decentralised precision agriculture and environmental monitoring.

This book also examines automated crop disease diagnosis using computer vision, reinforcement learning, soil health surveillance systems for quality improvement, agricultural data normalisation and integration for crop knowledge discovery. Researchers, academics, business professionals, legislators, and graduate students studying artificial intelligence, agricultural engineering, environmental sciences, and smart farming technologies are the target audience for this book. It aims to motivate the upcoming generation of climate-smart solutions by fusing applied creativity with academic rigour. In the end, Climate Smart Agriculture imagines a future in which farmers, ecosystems, and communities work together with intelligent systems to improve productivity, ensure sustainability, and fortify resilience against climate uncertainty. Our goal is for this work to be an academic resource as well as a catalyst for revolutionary change in the world of agriculture.

# 1

# Deep Reinforcement Learning from the Perspectives of Artificial Intelligence and Optimal Control

**J. Jesy Janet Kumari[1*], Thangam S.[2], Raghu Ramamoorthy[1] and Anitha Velu[3]**

*[1]Department of Computer Science and Engineering, The Oxford College of Engineering, Bengaluru, Visvesvaraya Technological University, Karnataka, India*
*[2]Department of Computer Science and Engineering, Amrita School of Computing, Amrita Vishwa Vidyapeetham, Bengaluru, India*
*[3]Department of Electronics and Communication Engineering, Sri Sairam College of Engineering, Bengaluru, Karnataka, India*

***Abstract***

Deep reinforcement learning (DRL) has become a viable representation learning technique in a range of machine learning (ML) courses, particularly in a reinforcement learning (RL) space, because of continuous technological breakthroughs addressed through artificial intelligence (AI). The recent trajectory has enabled the emergence of a novel technological domain known as DRL. When DRL agents needed to communicate in more complex, data-rich environments, the approach proved to be highly data-inefficient. The fundamental reason is the limited applicability of DRL techniques to a wide range of scenarios that correspond to connected tasks from an identical distribution. The application of a multitask instructional strategy might mitigate this challenge. This chapter presents an innovative, multitask, learning-based approach to enhance DRL agents operating in various semantically similar environments with comparable tasks. This section utilizes a global network that will enhance knowledge transmission among individual actor-critic models functioning in separate environments. Various agricultural challenges have been mitigated by applying AI technology, including diagnosing plant diseases, predicting crop yields, and identifying weeds. In contrast, not all AI models, such as DL and ML, exhibit comparable efficacy across a diverse array of agricultural datasets.

**Corresponding author*: ebijesy.j@gmail.com; ORCID: 0000-0002-3318-1632

Anitha Velu, Prasanth Aruchamy, Raghu Ramamoorthy, Rajesh Kumar Dhanaraj and Seifedine Kadry (eds.) Climate Smart Agriculture, (1–16) 

***Keywords***: Deep reinforcement learning, artificial intelligence, smart agriculture, deep learning optimal control, machine learning

## 1.1 Smart Agriculture

Among the world's largest farming countries is India. It grows different crops on soils and temperatures that differ from one another. Its produce consists of wheat, rice, lentils, fruits, vegetables, and spices. For millions of farmers, protecting food security and advancing rural development, India's agriculture sector [1] is absolutely vital. Moreover, challenges including water shortages, climate change, and land degradation face this industry, which calls for sustainable farming methods [2] and legislative interventions. By improving crop yields, reducing waste, and optimizing agricultural operations, machine learning (ML) can revolutionize the agriculture industry. By assessing data from sensors and satellite photos, ML can improve precision agriculture. This will let farmers decide which sections of their fields call for fertilizers, water, or other resources. Through crop health monitoring, farmers can examine nutrient deficits or parasite infestations and carry out preventative actions before the problem becomes more serious. Machine learning models can forecast future crop yields by means of past data on soil, climate, land use, and other crop yields. For different crops, this knowledge can help growers decide when to sow, pick, weed, and apply fertilizer.

### 1.1.1 Necessity of Smart Agriculture

Smart agriculture, also known as customized farming or smart farming, is a modern approach that uses data-driven solutions and advanced technologies to improve agricultural processes, increase crop growth [3], and decrease environmental effects. The fast expansion of the global population has made it necessary to modify conventional farming methods, use more intelligent and sustainable food production techniques, address issues related to high temperatures and limited resources, and change the desires of consumers.

## 1.2 Necessity for Deep Reinforcement Learning in Smart Agriculture

Machine learning can transform the agricultural industry by improving its profitability, sustainability, and yield. Agriculture is a highly influential contributing factor to the economy of any nation. Crop yield prediction

(CYP) [4] holds immense potential in understanding economics, supply, and demand. Consequently, when CYP is associated with an economy, it becomes strategic. Combining information with the agricultural cycle and environmental factors enhances the effectiveness of CYP. Factors, such as crop variety, field size/demography, growth process, soil type, humidity, temperature [5], the affordability of agricultural resources, and the models employed for prediction, improve the forecast's accuracy. The application of DL, ML, computational intelligence, and statistical simulation is a commonly utilized method for CYP. In the prior time frame, deep learning (DL) strategies were predominantly shaped by the accessibility of computational hardware and data sets. The accessibility of general processing units at affordable costs and their substantial estimation power has prompted developers to implement larger and deeper networks for CYP. When it comes to DL, diverse input samples are utilized to train the model. Weights are modified in accordance with the error value computed in every stage, reflecting changes made to each dimension in an advanced neural network model. Therefore, the models acquire the weight and use it to address classification, regression, and prediction issues.

Figure 1.1 shows the organizational structure of artificial intelligence (AI), ML, and deep reinforcement learning (DRL) systems. Several notable scientists and specialists in contemporary agriculture are conducting a more thorough examination of their hypothesis, which is helping to make more accurate and consistent predictions. Currently, smart agriculture utilizes different methods of ML to preserve rainwater and optimize the use of additives and energy in response to diverse environmental changes.

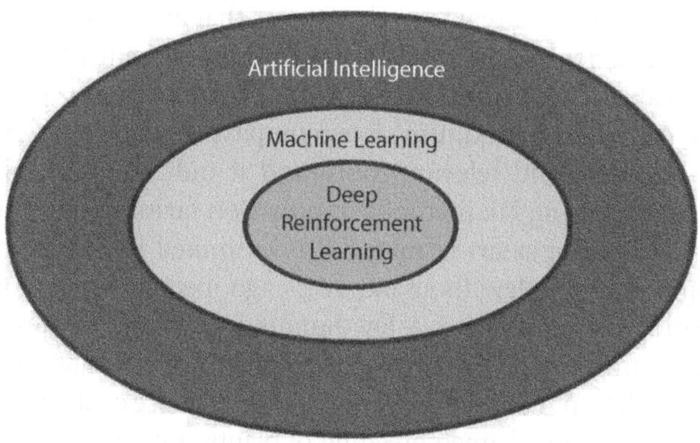


**Figure 1.1** The organizational structure of artificial intelligence, ML, and deep reinforcement learning.

Machine learning uses for agriculture enhance crop management through harvest forecasting, crop disease identification, weed detection, knowledge of plant varieties, and crop cycle prediction, among other functions [6]. In CYP, various algorithm taxonomies utilize a range of essential features. The analysis incorporates variables such as temperature, precipitation, and soil type, utilizing algorithms such as convolutional neural networks (CNNs), long short-term memory (LSTM), and deep neural network (DNN) techniques, with a forecast accuracy of 92%.

### 1.2.1 Importance of Reinforcement Learning

Reinforcement learning (RL) is a technique utilized to solve problems that involve sequentially making decisions. The rapid development of DNNs has facilitated notable progress in RL in the past few years. Reinforcement learning has the potential to be applied in various fields, including robotics and gaming. Conversely, transfer learning has surfaced as a remedy for the issues that RL faces. Applying knowledge enhances the effectiveness and efficiency of the learning process. IT describes the comprehensive analysis of the significant advancements in transfer approaches to acquiring knowledge through DRL. Agri-Internet of Technology (IoT) [7] serves as a conceptual foundation for IoT-based intelligent agricultural applications. This section provides on-demand information from the sensors. It offers an extensive semantic processing pipeline that integrates various cross-domain data streams, providing an integrated structure for smart applications related to agriculture. Agri-IoT facilitates extensive analysis of data and event detection, ensuring the smooth compatibility of indicators, facilities, operations, cultivators, and other important participants, as well as online data sources, streams, and statistics that are integrated and easily accessible. In agriculture, the AI-based training system [8] conducted a thorough evaluation of research focused on the utilization of DNN technology in precision agriculture. It encompasses real-time applications, accessible datasets, and relevant tools, and it indicates the considerable potential of employing DL methods in precision farming systems.

The significance of smart farming [9] is examined in the context of sustainable agriculture. A few thousand years ago, people tamed animals and plants, turning them into pets. A few hundred years ago, crop rotation and other farming improvements were used regularly. And just a few decades ago, the "green revolution" happened with the widespread use of artificial fertilizers and pesticides and the planned breeding of plants and animals. It shows that the steadily growing prevalence of information and communication technology in agriculture is the catalyst for the fourth agricultural revolution.

To establish a rapid and reliable agricultural production prediction system, a new approach must be used to predict fruit production using DNNs [10]. It analyzes various fruit production data, including apples, bananas, citrus, pears, grapes, and overall fruits, and uses DNNs to predict the future evolution of these fruits. The data pertains to the production output of primary fruits and comes from Pakistan's National Bureau of Statistics. Deep neural networks present a specific methodology to project future production data, such as Levenberg–Marquardt optimization, which attained an accuracy of 65.6%.

Clustering algorithms [11] were applied to the Food and Agriculture Organization soybean dataset, and a performance analysis was conducted. Based on certain parameters, the experiment is conducted using various parameters, such as algorithmic complexity, iteration count, cluster formation, and completion time. It determines the most effective blending method for the Food and Agriculture Organization soybean dataset *via* analysis and reviews neural network research in agricultural production systems. Crop management involves forecasting yields, identifying diseases, weeds, and crop quality, as well as managing the harvesting systems. Categories were assigned to analyze works. The pros and cons of DL and future research were listed [12]. Deep learning research is suggested as more accurate than ML, and also compared with Caffe, Neon, TensorFlow, Theano, and Torch on criteria such as flexibility, hardware application, and performance, as well as it evaluates the effectiveness of the frameworks on a single machine utilizing both the central processing unit multi-threaded and graphics processing unit Nvidia Titan X configurations and it examines various DL architectures. It also considers performance metrics, including the gradient time for computation during the training process for deep networks, and the forward time during the implementation of trained networks. It provides information on the performance of various convolutional algorithms and the support provided by all these frameworks for convolutional networks. The chapter offers a framework for classifying the most advanced transfer learning approaches. It is also used to evaluate the objectives, methodologies, use cases, and corresponding RL backbones of these approaches. Reinforcement learning is an effective framework for solving sequential decision-making tasks. In this section, an agent learns how to communicate with its surroundings to enhance its ability to learn through trial and error [13]. Emerging from the field of cybernetics, RL has flourished in computer science, addressing previously insurmountable challenges.

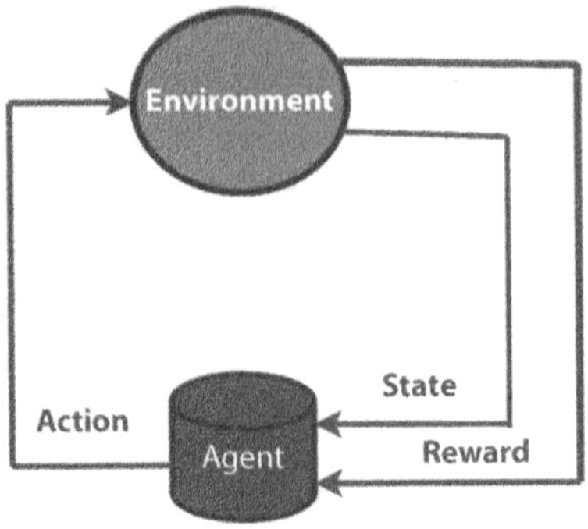


**Figure 1.2** Working environment of DRL.

Figure 1.2 describes the working environment of DRL. The development of conventional RL algorithms primarily focused on tabular cases, providing principled solutions for simple tasks. However, they encounter challenges when managing complex domains, such as tasks that involve 3D environments [14]. Recently developed DNNs and RL have been combined to tackle complex tasks. This combination is called DRL. This approach employs DNNs to develop robust function approximators that can effectively navigate complex domains. Deep reinforcement learning has obtained significant adoption in applications such as robot control and games. It also flourishes in a variety of domains, including smart transportation systems, electricity networks, and health informatics. Despite its remarkable progress, the exploration-exploitation problem continues to pose intriguing challenges for RL [15]. Real RL problems typically lack knowledge of environmental dynamics, and the agent cannot utilize this knowledge until it accumulates enough interaction experiences through exploration. In domains such as automatic driving and health informatics, the complex nature of state and action spaces, sparse feedback, and partial observability can make the acquisition of adequate interacting sampling prohibitive or even pose safety concerns. The challenges have prompted multiple efforts to improve the current RL procedure. Transferable Learning (TL) [14], or the transfer of knowledge, is a method that leverages external expertise to enhance learning in the target domain. Consequently, it is an essential subject in RL. Despite extensive research on TL techniques in controlled learning, RL is still a developing field.

Transferable learning in RL becomes increasingly complex within the Markov decision process framework, necessitating the transfer of knowledge. Furthermore, given the intricate components of the Markov decision process, expert knowledge may present itself in multiple forms that require different transfer methods. Previous efforts at summarizing TL in the RL domain overlooked research conducted over the past century, a period during which significant advancements in transfer learning were facilitated through advanced ML techniques.

### 1.2.2 Deep Learning Approaches

Reinforcement learning and DL are combined to form DRL. This approach has successfully handled numerous intricate decision-making tasks that were previously unsolvable by a computer. As a result, DRL has the potential to create a vast number of new applications in a variety of disciplines, including economics, machinery, healthcare, and smart infrastructure. It serves as an introduction to the models, algorithms, and techniques of DRL. Sequential decision-making is the primary focus of RL, a branch of ML; it formalizes the RL problem in the form of an agent that is required to make decisions in this chapter and the environment to optimize a specific concept of progressive benefits. It will be clear that this formalization covers many important aspects of AI [16], such as the perception of cause and consequence, ambiguity, and nondeterminism, and that it applies to a wide range of tasks. Additionally, it provides a summary of the many approaches for learning consecutive decision-making processes and the possible advantages of deep learning library (DLL). The DLL computational methods used in the optimal control solution are presented in Figure 1.3.

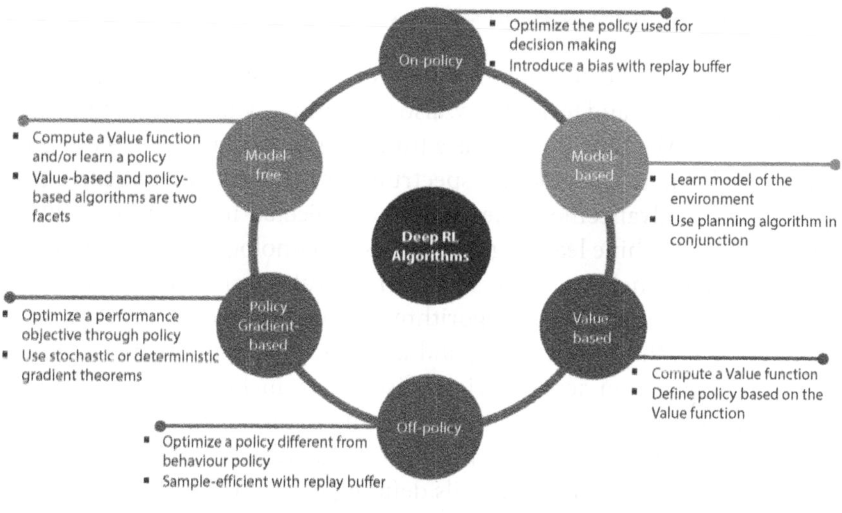


**Figure 1.3** Algorithms of deep DRL in optimal control solution.

## 1.3 Machine Learning

The branch of ML [17] known as DLL combines two distinct approaches, namely RL [18] and DL. It gives its agents the ability to make decisions based on unorganized data. If a significant volume of information needs to be processed, it can be utilized in certain situations. An intermittent DL model developed on top of the RL algorithm is known as the deep recurrent Q-network model.

### 1.3.1 The Need for Artificial Neural Networks

Inspired by biological neural networks, artificial neural networks (ANNs) [19] are a subset of neural networks. They are usually constructed using artificial neurons, which are theoretically similar to biological neurons. The neurons in a network of artificial neurons process input and produce output signals in reaction to specific stimuli. In general, the computational techniques that make up an ANN are responsible for problem-solving. They respond similarly to how people respond. They stimulate the signals in the same way as neurons do in the human brain. Crop yield estimation for harvests is one of the many uses of ANNs in agriculture. It was discovered that the accuracy of the ANN calculation was higher than that of the support vector machine calculation when the two methods were applied to binary classification. In this context, researchers studied the combination of ANN and CNN to identify leaf diseases and recommend pesticides.

### 1.3.2 Intelligent Artificial System

Artificial intelligence computation has become a significant topic in contemporary science and technology, also indicating a vital domain for future development. Machine learning is a fundamental technology in computational intelligence. The practical spectrum for ML is broadening alongside technological advancements, leading to a projected increase in its future applications. Machine learning constitutes a technology grounded in intelligence-based principles. The system extracts rules from extensive datasets and continually improves its algorithmic model to predict and categorize unknown data through learning and analysis. Machine learning technology finds application across various disciplines, including intelligent recommendation systems, computer vision, and natural language processing.

The development of ML methods can be divided into three phases. In the first step, rule-based ML entails defining rules for prediction and data classification. The second phase consists of statistical data-dependent ML.

By means of statistical analysis, this ML technique generates information that helps identify and classify previously unidentified data. The core of the upcoming stage of ML is DL. By recreating the framework and capabilities humans offer neural networks, this ML approach helps with data analysis and prediction. Machine learning methods are used in many fields, including natural language processing, which depends heavily on them. By learning and analyzing a significant volume of natural language data, the ML system completes speech recognition, language comprehension, language conversion, and other tasks. One important use of ML technologies is computer vision.

Recognizing photos, identifying faces, and tracking objectives are just a few of the tasks the ML system can handle. It learns and analyzes data on video and picture sources to accomplish this. Another use of ML technology is intelligent recommendation. By analyzing users' past behavior and preferences, the ML system can offer customized recommendations, buy suggestions, and other capabilities. The control of financial risk is among the most important uses for AI technology. Gathering and analyzing enormous volumes of banking data allows an ML-based system to perform numerous functions like credit assessment and fraud detection. Effective ML systems depend on the regulated extraction of data-based rules and patterns. Many jobs, including classification, decision-making, and prediction-making, call for their application. Learning methods are based on neural networks; DL is a subset of ML methods because this system can independently identify complex patterns and traits from large datasets, benefiting many disciplines, including voice and image recognition. Natural language processing is a new technology that converts human conversation into a format that computers can understand.

This section applies to emotion analysis, machine translation, and text classification. Data mining is a method for identifying implicit knowledge and norms based on vast data quantities. This knowledge can be utilized for several purposes, such as categorization, prediction, and association analysis. Figure 1.3 illustrates the fundamental framework of general artificial intelligence.

The estimation of crop yield utilizing ecological, soil, water, and crop factors presents a viable research topic. Deep-learning-based models are commonly applied to extract important crop features for predictive purposes. These methods may address the yield prediction issue; however, they exhibit the following shortcomings: a direct non-linear or proportional connection between the unprocessed information and crop yield values cannot be established. The accuracy of the extracted characteristics significantly influences the models' efficacy. Deep reinforcement learning offers guidance and impetus for addressing deficiencies; it serves as a robust framework for CYP, effectively transforming raw data into crop prediction values by integrating RL and DL methodologies.

To estimate agricultural output, it creates a deep recurrent Q-network model that combines the Q-learning RL algorithm with a recurrent neural network (RNN) algorithm for DL. The successive layers of the RNN process the data parameters. Based on parameters, the Q-learning network creates an environment for CYP. A linear layer converts the resulting RNN principles into Q-values. An RL agent uses a restricted and parametric feature for CYP.

Finally, the lowest possible error and the highest possible forecast accuracy determine the agent's aggregate score for their actions. With an accuracy of 93.7%, the model outperforms extant models in CYP by preserving the original data distribution.

## 1.4 Applications of Deep Learning in Smart Agriculture

Deep learning is the most advanced ML technology, demonstrating exceptional performance in a variety of fields, including bioinformatics, language processing, and computer vision. Agricultural research extensively employs DL due to its exceptional image processing capabilities. Plant or crop classification constitutes most DL applications in agriculture. This classification is essential for a variety of purposes, including pest control, autonomous harvesting, yield prediction, and disaster monitoring.

Manually detecting plant diseases is time-consuming. Fortunately, advancing DL enabled the detection of plant diseases through image processing. Examples of CNN architectures that can identify plant leaf maladies include AlexNet, GoogLeNet, and ResNet. Deep neural networks can also address the critical issue of meteorological forecasting in agriculture. Predicting crop yields before harvest helps landowners, customers, and the government plan for market intervention, alleviating food shortages, marketing, and procurement. Conditional restricted Boltzmann machines and CNNs are two DL methods that provide accurate representations, classification, and predict rainfall with sufficient precision.

### 1.4.1 Land Cover Classification

Land cover classification, a critical and challenging task in agriculture, implements DL with the primary objective of classifying a typical piece of land. Methods such as CNN, generative adversarial network, and RNN can classify satellite image data into land cover. In agriculture, to monitor the health of crops and soils, companies are utilizing ML and DL techniques to process data collected by unmanned aircraft and/or software-based technologies. Researchers also use DL to monitor and anticipate a variety of environmental factors, such as weather fluctuations that affect crop yield. In these kinds of reviews, the other two most often used DL techniques are DNN and LSTM. The most used DL algorithms are CNNs.

### 1.4.2 Convolutional Neural Network

Methods of DL applied in CNN, RNN, and VGGnet in the visual examination of leaves to identify crop diseases using image-based anomaly detection solutions. The method developed in the data envelopment analysis (DEA) uses eight physiological signals and a CNN model to detect anomalies in physiological signals. Deep learning algorithms analyze crop performance in a variety of climates and develop new characteristics as they process decades of field data. Based on this data, they can construct a probability model that anticipates which alleles are most likely to contribute a beneficial trait in a plant. Deep learning algorithms, including RNNs, LSTMs, autoencoders, deep Boltzmann machines, and deep belief networks, are utilized to monitor soil moisture and temperature, facilitating an understanding of ecosystem dynamics and their implications for agriculture. The relationship between DL algorithms, including feedforward neural networks, CNNs, and RNNs, and the processing of daily, weekly, or monthly data, is significant. Transpiration enables forecasting of daily dewpoint temperatures and the more effective use of irrigation systems. This connection helps identify anticipated weather phenomena and estimate transpiration and evaporation. By providing precise predictions and measurements of farming parameters, DL methods, such as CNN, RNN, and LSTM, can maximize the economic efficiency of livestock production systems, including the production of eggs and cattle.

## 1.5 Impact of Deep Reinforcement Learning on Artificial Intelligence and Optimal Control

Deep reinforcement learning uses neural networks. Although the function that a network approximates varies based on the technique, the procedures for training networks have many similarities, including loss calculation and parameter updates, and it is beneficial to adopt a consistent net application programming interface (API) to access all neural network structures employed by a variety of algorithms. Utilizing modularized network components makes it easier to read and debug algorithm implementations since they require less code.

The following requirements must be met to create a Net API for deep RL:

### 1.5.1 Input and Output

The changes are based on the combination of the environment and the algorithm. Users do not need to manually enter their input and output dimensions at all times because these can be inferred automatically. This functionality saves time and reduces the likelihood of coding errors.

### 1.5.2 Automatic Network Construction

The design of a network significantly influences how well an algorithm performs, and it differs depending on the environment. Because it is typical to experiment with a variety of network architectures in DRL, it is preferable to describe network architectures *via* a configuration file rather than changing the code. This necessitates a technique for taking a setup and automatically building the neural network that corresponds to it.

### 1.5.3 Training Step

Every neural network requires a training step that consists of calculating the loss, determining the gradient, and modifying the parameters of the network. Standardizing these processes into a single function is beneficial because it allows them to be reused for every algorithm.

### 1.5.4 Visibility of Fundamental Methods

The API, which sits on top of neural network libraries, must additionally disclose its most frequently utilized techniques, including activation

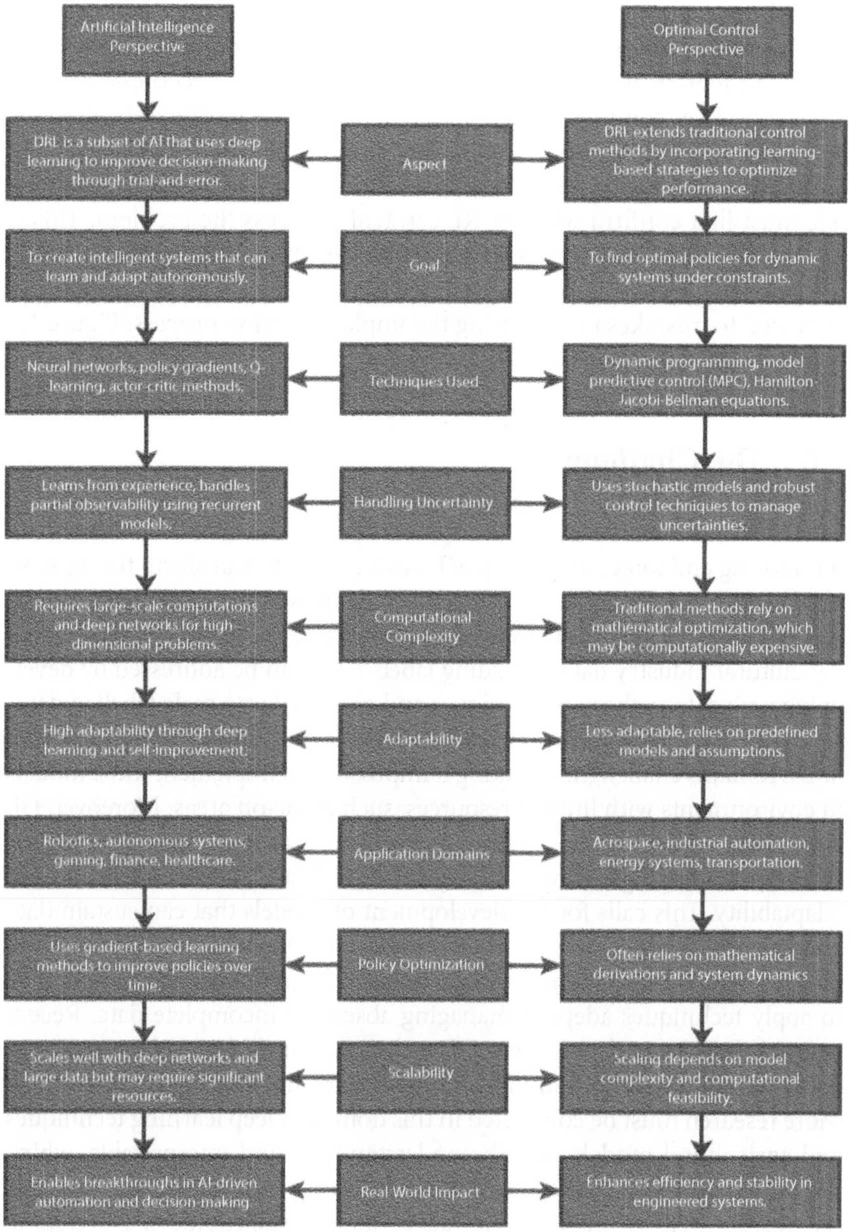


**Figure 1.4** Aspects of DRL.

algorithms, optimization techniques, learning rate decay, and network checkpointing. One of the most significant considerations is the distinction between theoretical correctness and implementation correctness. A novel RL method or component must first demonstrate its theoretical validity before implementation. This is very important when doing research. Similarly, before using any methods to address unfamiliar circumstances, one must first confirm whether RL can truly address the problem. This is particularly important for applications. If everything is theoretically accurate and the problem can be solved, the malfunction of an RL system can be traced to mistakes made during the implementation process. Figure 1.4 below shows the aspects of a DRL network.

## 1.6 The Challenges

By facilitating increased cultivation, precision agriculture, and enhanced crop monitoring and forecasting, deep ML has a chance to transform the agricultural industry. To overcome the several challenges in maximizing the possibilities of ML in agriculture, one major challenge is the lack of exceptional agricultural industry data, including labels. This can be addressed by developing new data-gathering techniques and creating large-scale labelled datasets to feed DL algorithms. Furthermore, the significant computational cost of DL remains challenging, making it impossible to implement these models in environments with limited resources, such as far-off areas. Moreover, DL algorithms applied in agriculture must adapt to a range of environments, crop types, imaging settings, and sensor modalities to reach resilience and adaptability. This calls for the development of models that can sustain data and generally apply across several situations. Moreover, as agricultural data usually shows incompleteness, noise, or corruption, it is imperative to apply techniques adept at managing absent or incomplete data. Recent research has focused on strong DL techniques, including robustness augmentation, adversarial training, and meta-learning, to solve this challenge. More research must be conducted in this domain. Deep learning techniques and agricultural models must be understandable and interpretable, which presents one major challenge. Making decisions with knowledge depends on this, as well as building confidence among stakeholders. Developing methods that improve the comprehension of the foundations of DL model processes becomes more important. Additionally, there are interpretable DL methods, including rule-based systems and decision trees. Including appearance detectors and weather data, among other data types, helps DL methods be more effective in agriculture.

## 1.7 Conclusions and Future Scope

This chapter presented a thorough review of the latest developments in the agricultural sector linked to DL, as well as a thorough study of DL in agricultural approaches. The main goal was to do this utilizing several DRL methods. Although DL methods have been implemented in agriculture, this technology continues to face challenges. The challenges encompass the necessity for DL experts to guarantee high production, the duration needed for employee training, and the aggregation of datasets. The efficiency of the system may be influenced by the implementation of large-scale modelling and software on small devices, such as smartphones, as well as the establishment of structures and hardware requirements. The application of DL methods, as highlighted by recent predictive modelling and tools, has advanced the agriculture sector. The DL approach demonstrates considerable potential within the agricultural sector. The high costs associated with hardware and software significantly limit the implementation of this technology. Therefore, conducting future research and development is essential to enhance the efficiency of large-scale applications through cost-effective DL techniques. In addition, there are restrictions on the precision and efficacy of ML algorithms. As a result, additional study is needed in this area to maximize the accuracy and efficiency of DL techniques. The accessibility of computational resources hinders the deployment of sophisticated ML techniques in agriculture. The complex nature of DL strategies, neural networks in particular, and the expanding volume of datasets require a continually increasing allocation of computing resources for model development. The accessibility of cloud computing services, particularly the use of Google Cloud Platform by small firms, has accelerated the development of DL. Relevant uses for DL in agriculture predominantly function offline, owing to stringent computational capacity requirements. This suggests a non-correlated method for the collection and examination of agricultural metaphors.

**AI disclosure statement:** The author confirms that no artificial intelligence (AI)-generated content or AI-assisted writing tools were used in the preparation of this chapter (including summaries and the abstract). All text, analysis, and conclusions are the result of the author's original work.

## References

1. Albahar, M., A survey on deep learning and its impact on agriculture: Challenges and opportunities. *J. Agric.*, 13, 3, 2023.

2. Altalak, M., Ammad uddin, M., Alajmi, A., Rizg, A., Smart agriculture applications using deep learning technologies: A survey. *Appl. Sci.*, 12, 5919, 2022.
3. Cantini, R., Marozzo, F., Orsino, A., Deep learning meets smart agriculture: Using LSTM networks to handle anomalous and missing sensor data in the compute continuum, in: *Device-Edge-Cloud Continuum: Paradigms, Architectures and Applications*, pp. 141–153, Springer Nature Switzerland, Cham, 2023.
4. Wang, X., Wang, S., Liang, X., Zhao, D., Huang, J., Xu, X., Miao, Q., Deep reinforcement learning: A survey. *IEEE Trans. Neural Networks*, 35, 5064, 2022.
5. Arulkumaran, K., Deisenroth, M.P., Brundage, M., Bharath, A.A., A brief survey of deep reinforcement learning, arXiv preprint arXiv:1708.05866, 2017.
6. Gao, Q. and Schweidtmann, A.M., Deep reinforcement learning for process design: Review and perspective. *Curr. Opin. Chem. Eng.*, 44, 101012, 2024.
7. Lazaridis, A., Fachantidis, A., Vlahavas, I., Deep reinforcement learning: A state-of-the-art walkthrough. *J. Artif. Intell.*, 69, 1421, 2020.
8. Shuford, J., Deep reinforcement learning unleashing the power of AI in decision-making. *J. Artif. Intell. Gen. Sci.*, 1, 4023, 2024.
9. Millea, A., Deep reinforcement learning for trading - A critical survey. *Data*, 6, 119, 2021.
10. Treloar, N.J., Braniff, N., Ingalls, B., Barnes, C.P., Deep reinforcement learning for optimal experimental design in biology. *PLoS Comput. Biol.*, 18, e1010695, 2022.
11. Stranieri, F., Stella, F., Kouki, C., Performance of deep reinforcement learning algorithms in two-echelon inventory control systems. *Int. J. Prod. Res.*, 1, 16, 2024.
12. Rolf, B., Jackson, I., Müller, M., Lang, S., Reggelin, T., Ivanov, D., A review on reinforcement learning algorithms and applications in supply chain management. *Int. J. Prod. Res.*, 61, 7151, 2023.
13. Samtani, S., Zhu, H., Padmanabhan, B., Chai, Y., Chen, H., Nunamaker Jr., J.F., Deep learning for information systems research. *J. Manag. Inf. Syst.*, 40, 271, 2023.
14. Mei, Z., Yang, Z., Chen, J., A novel reinforcement learning-based method for structure optimization. *Eng. Optim.*, 1, 20, 2024.
15. Panzer, M. and Bender, B., Deep reinforcement learning in production systems: A systematic literature review. *Int. J. Prod. Res.*, 60, 4316, 2022.
16. Li, Y., Deep Reinforcement Learning: An Overview, arXiv preprint arXiv:1701.07274, 2017.
17. Rabault, J., Kuchta, M., Jensen, A., Réglade, U., Cerardi, N., Artificial neural networks trained through deep reinforcement learning discover control strategies for active flow control. *J. Fluid Mech.*, 865, 281, 2019.
18. Liu, Z. and Mu, S.U., An overview on algorithms and applications of deep reinforcement learning. *Chin. J. Intell. Sci. Technol.*, 2, 314, 2020.
19. Hu, W. and Hu, J., Reinforcement learning with deep quantum neural networks. *J. Quantum Inf. Sci.*, 9, 14, 2019.

# 2

# Climate-Smart Agriculture: Adoption, Impacts, and Implications for Sustainable Development

**J. Chandra Priya[1]*, G. Nanthakumar[2], C. Alamelu[3] and Afizan Bin Azman[4]**

[1]*Department of Computer Science and Engineering, National Institute of Technology, Tiruchirappalli, Tamil Nadu, India*
[2]*Department of Artificial Intelligence and Data Science, Anjalai Ammal Mahalingam Engineering College, Tiruvarur, Tamil Nadu, India*
[3]*Department of Computer Science and Engineering, Amrita School of Computing, Amrita Vishwa Vidyapeettham, Chennai, Tamil Nadu, India*
[4]*Faculty of Innovation & Technology, Taylors University, Subang Jaya, Selangor, Malaysia*

***Abstract***

This chapter explores the adoption, impacts, and implications of climate-smart agriculture (CSA) for sustainable development. Climate-smart agriculture is an integrated approach that enhances agricultural productivity while adapting to climate variabilities and mitigating greenhouse gas (GHG) emissions. The technologies involved in CSA are analyzed in key areas, such as precision agriculture technologies and data analytics, to optimize field-level management regarding crop farming. The integration of artificial intelligence (AI), the Internet of Things (IoT), and remote monitoring technologies enables real-time monitoring of agricultural conditions, facilitating timely actions. Precision irrigation systems optimize water use efficiency, and applying data analytics enhances fertilizer application efficiency, thereby reducing waste and mitigating environmental impacts. Agroforestry and conservation agriculture techniques, such as no-till farming and cover cropping, contribute to improved biodiversity and enhanced soil resilience in the face of climate variability. Additionally, CSA emphasizes the utilization of climate-resilient crops, integrated pest management strategies for ecological safety, renewable energy solutions, and efficient livestock management systems.

**Corresponding author*: jchandrapriya@gmail.com; ORCID: 0000-0003-3891-0950

Anitha Velu, Prasanth Aruchamy, Raghu Ramamoorthy, Rajesh Kumar Dhanaraj and Seifedine Kadry (eds.) Climate Smart Agriculture, (17–40) © 2026 Scrivener Publishing LLC

***Keywords***: Data-driven agriculture, sustainable agriculture, precision farming, climate-resilient crops, crop cycling

## 2.1 Overview of Climate Change Impacts on Agriculture

The Indian agricultural sector is highly dependent on natural environmental conditions and faces significant challenges due to climate change [1]. Abnormal climate variations can drastically affect crop productivity, leading to failures and altering agricultural production modes. Climate change also influences water resources, which, in turn, affects agricultural productivity [2]. Conversely, the agriculture sector is a significant source of greenhouse gas (GHG) emissions that lead to climate change. According to the Intergovernmental Panel on Climate Change, agriculture accounts for approximately 23–25% of anthropogenic GHG emissions globally. Carbon dioxide ($CO_2$) is emitted mainly from the microbial breakdown of organic waste materials in the soil and the ignition of excess agricultural residues. Methane ($CH_4$) emissions are from anaerobic soil in paddy fields, anaerobic microbial activities in the gastrointestinal tract of cattle, and organic matter decomposition in manure and crop wastes under waterlogged environments [3–5]. The causes of nitrous oxide ($N_2O$) pollution include nitrogen-rich soil, animal waste, and composted materials.

Agriculture contributes to GHG emissions; however, it also mitigates the impacts of climate variability. Climate change worsens existing challenges in agriculture, resulting in low yields due to droughts, heat waves, and floods. Hence, it necessitates farmers to reassess and improve their farming systems, involving a paradigm shift in processes, practices, and structures, inspired by the necessity of change that climate-resilient farming methods can induce [6, 7]. However, the adoption of CSA varies by demographic, economic, and cultural factors. While climate-resilient farming methods, such as conservation agriculture and agroforestry, have received more support, adoption rates are low globally. If CSA practices include a shift in crop patterns, planting dates, and water-saving strategies, productivity and labor returns also rise [8]. Practices such as harvest seasons, food scarcity management, and soil erosion mitigation, used alone or in combination with CSA technologies, can alleviate the consequences of climate variability for farming systems [9]. With attention to energy efficiency and its implications on farm profits, and ecological issues, research by some authors reiterates the merits of CSA methods in terms of ecological integrity, economy, and social welfare [10].

Figure 2.1 depicts the CSA methods practiced in many countries to improve agricultural sustainability and solve issues unique to a given area. While Namibia uses fog water collecting and hydroponic farming to adapt to arid conditions, Malawi uses agroforestry systems and continuous intercropping to increase biodiversity and productivity. Countries like China focus more on raising the volume of soil organic carbon while reducing $N_2O$ and methane ($CH_4$) emissions. The reuse of straws enhances soil fertility to improve microbial activities, which is what Nepal has been following. In countries like Vietnam, agricultural productivity is achieved by reducing environmental impacts by altering the plantation timings and minimizing the use of chemical fertilizers. This practice promotes sustainability in farming and protects local ecosystems, which supports long-term food productivity and security. Water scarcity in Pakistan is enhanced through wastewater management, implementing water pricing models, and adopting drought-resistant crop varieties. Countries such as Zambia progress in sustainable agriculture through crop rotation, organic mulching, and agroforestry. Additionally, conservation farming techniques, such as precision irrigation techniques with drip irrigation and micro-sprinkler systems, maximize water management in the US. It is inferred that these region-specific strategies project the global adaptation in CSA, which contributes to sustainability in agricultural production, resource management, and climate resilience [11].

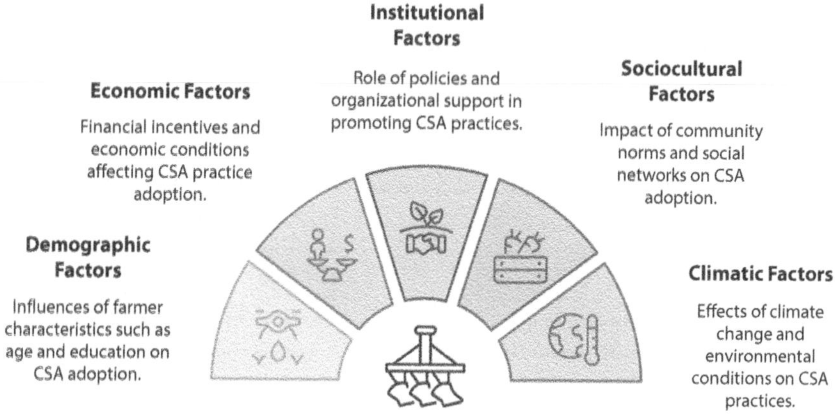


**Figure 2.1** Factors influencing CSA adoption.

In a highly populated country like India, CSA has emerged as an essential approach for enhancing food security and modifying practices to climate dynamics. The conceptual model of CSA was initiated by the Food and Agriculture Organization of the United Nations during the 2010 Hague Conference. The three primary objectives proposed at the conference include lowering GHG emissions, enhancing the resilience of farming systems to the impacts of climate change, and sustainably boosting agricultural productivity. To accomplish these objectives, the Indian farmers achieve higher yields with reduced environmental impact by adopting improved crop varieties, implementing effective water management practices, and promoting sustainable soil health strategies [12, 13]. The region-specific localized solutions consider climatic and socioeconomic factors to improve farming. Figure 2.2 highlights three fundamental components such as sustainably increasing agricultural outcomes, mitigating vulnerability through climate-resilient adaptation measures, and reducing GHG emissions. It forms the foundation of CSA initiatives globally. In line with these observations, the primary objective of this article is a

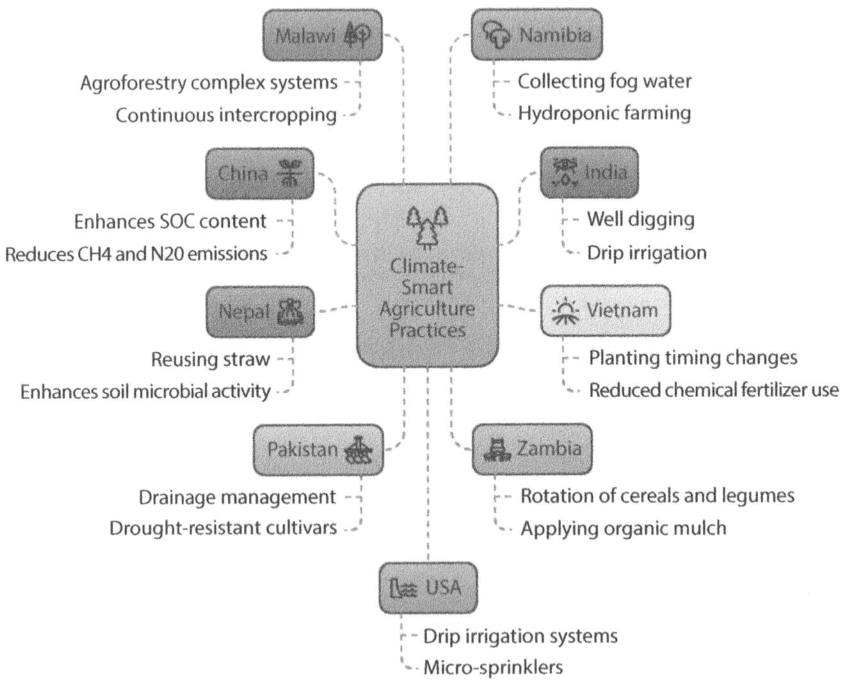


**Figure 2.2** Global CSA strategies.

comprehensive analysis of applying modern solutions, such as artificial intelligence (AI) and the Internet of Things (IoT), in CSA to increase productivity and enhance sustainable outcomes. The sub-objectives include,

- To evaluate the vulnerability to climate change and the resilience of food security among various livelihood groups in climate-vulnerable regions.
- Analyzing precision agriculture with remote sensing and water management in improving CSA practices for real-time monitoring and improved agricultural management.
- Exploring agroforestry and renewable energy that contribute to environmental sustainability and climate adaptation.
- Evaluating how data analytics techniques provide data-driven insights for crop farming with a focus on optimizing resource usage, such as water and nutrients, thereby minimizing waste and reducing environmental impact.
- Outlining a comprehensive plan for the integration of AI and IoT technologies that emphasizes custom-configured, context-specific solutions that improve productivity, promote sustainability, and increase environmental resilience.

## 2.2 Mixed-Method Approach to Climate-Smart Agriculture

This chapter combines a mixed-method research design with quantitative and qualitative approaches to present an in-depth analysis of the role of modern technologies in CSA. This methodology is designed to evaluate the adaptation, implementation, and effects of CSA on both agricultural productivity and environmental sustainability. It addresses three main research queries: (1) How can AI and IoT be integrated into precision agriculture to improve real-time monitoring and management practices in climate-vulnerable regions? (2) What are the effects of advanced water management techniques like precision irrigation on improving water-use efficiency and crop productivity within the CSA framework? and (3) How do agroforestry and renewable energy contribute to biodiversity, soil health, and sustainability in climate-impacted agricultural systems? To answer these research questions, an extensive literature review was conducted using the Scopus and Web of Science databases, targeting publications from 2015 to 2024. The search used relevant keywords combined with Boolean operators to retrieve the most relevant articles. Initially, 28 articles were identified and

**Table 2.1** The formulation and strategy for search queries.

| Search query | Search keywords | Scopus results | Science direct results | IEEE xplore results |
|---|---|---|---|---|
| Q1 | "Climate Smart Agriculture" AND "Artificial Intelligence" | 20 | 15 | 5 |
| Q2 | "Precision Agriculture" AND "IoT" | 15 | 10 | 3 |
| Q3 | "Agroforestry" AND "Sustainability" | 12 | 8 | 2 |
| Q4 | "Water Management" AND "Data Analytics" | 10 | 6 | 2 |
| Q5 | "Renewable Energy" AND "Crop Management" | 11 | 7 | 1 |

reviewed, with conference papers excluded to maintain the highest quality research. The search strategy is detailed in Table 2.1, listing the specific keywords used for each query and the number of results correspondingly retrieved from Scopus, ScienceDirect, and IEEE Xplore.

## 2.3 Multi-Stakeholder Technological Intervention Model

The conceptual framework of CSA integrates modern technologies with traditional agricultural knowledge to address climate risks and improve farming outcomes. It prioritizes resilience, inclusivity, and efficiency, engaging diverse stakeholders, including commercial and smallholder farmers, pastoralists, and farm workers. The conceptual framework considers three analytical levels: (1) micro-level stimuli, such as long-term climate trends; (2) meso-level factors that include organizational and socioeconomic dynamics; and (3) macro-level phenomena, including floods and droughts. The advent of advanced technologies, such as precision watering, data science farming, and IoT-enabled real-time monitoring, enhances overall productivity and sustainability by exposing resistance to climate variations [14]. With these considerations, the CSA practices include strategic farming techniques to improve efficiency and adaptability, with environmental awareness. It mandates integrating novel technologies with localized knowledge and expertise.

### 2.3.1 Stakeholder Categories

Within the CSA framework, various stakeholders exist with different roles in promoting environmentally friendly farming techniques alongside weather predictions. Based on the framework, pastoralists and livestock owners are an important category of stakeholders because livestock accounts for up to 14.5% of anthropogenic GHG emissions. It can be achieved through enteric fermentation and management of manure. They are also contributing to the implementation of carbon-sequestration methods like rotational grazing and agroforestry. Mixed farming systems with crop-livestock integration improve nutrient cycling, which reduces feed costs. These incorporations positively impact the pasture recovery rates and water resource management. Also, the implementation of precision livestock technologies and early warning systems adds to CSA practices that lead to sustainable fodder production and manure recycling. The other category comprises smallholder farmers who produce one-third of global food. Hence, their participation contributes to local and global food security. Most of the underdeveloped countries rely on this category because they are the majority of food producers. This category focuses on practices such as intercropping, composting, and strategic tree planting to increase yield, restore soil health, and conserve water.

Another set of stakeholders is farm laborers, who are the most vulnerable in the CSA framework, to economic instability. They primarily focus on techniques including row planting, cover cropping, and precision irrigation to reduce emissions. With these efforts, CSA practices tend to minimize socioeconomic variations within the environment to promote a reasonable and strong agronomic framework for all contributors. But the commercial producers are mainly responsible for technological use and its evolution. They are the majority of the farming community who use the latest tools and practices to enhance the workflow and increase production. These are the type of farmers who can invest both time and money to allocate resources for novel solutions that improve resource utilization and increase outputs. As a result, this commercial type of farming leads to technical progress that promotes sustainability and climate resilience. These farmers aim to increase their productivity and set a benchmark for smallholder farmers by using new tools and technology to modernize the agricultural sector and be climate-adaptive.

### 2.3.2 Vulnerability Context

The CSA framework addresses various vulnerabilities across different dimensions. Each degree of vulnerability creates limitations on both resilience and agricultural outputs. From the macro level aspect, extreme intrusions such as floods, droughts, and intense heat lead to considerable threats. The extreme change in weather patterns affects global food security and economic livelihoods by reducing crop yields, changing soil structure, and affecting livestock mortality. As such adverse conditions become more frequent and severe, the agricultural systems become more vulnerable. Hence, the reliability and the output of food production are affected. The consequences of global warming cause a sudden temperature rise, disrupting pollinator-plant synchronization and threatening reproductive hit. On considering the meso-level, the Economic, Institutional, and Market (EIM) factors directly influence the decision of farmers to adapt and build resilience. The market accessibilities and resource availability through community-based efforts, such as Farmer Producer Organizations (FPOs) and financial assistance, can be extended. This kind of support to the farmers provides options for choosing alternatives such as rescheduling of sowing dates and crop densities to increase their income with better market prices.

Within the micro level of the framework, the majority of the local farmers face challenges due to the changing and unpredictable climate patterns. The climate variabilities such as variations in the seasonal rainfall, precipitation pattern changes, unusual temperature rise, and increased pests or disease incidents worsen the already existing vulnerabilities. Out of such, the common farm-level issues are decreased crop output, water resources, and soil nutrients. To cope up with these problems, most farmers are inclining towards adaptive measures like the use of crops planned for drought tolerance, improved soil management, and water conservation techniques. Beyond these challenges, the agricultural sector faces alarming threats from climate change caused by industrial emissions, water shortages, and rising GHG levels. Hence, coordinated efforts are required that include policy reforms for subsidies, incentivizing CSA practices and restoration of forestry ecosystems to augment carbon sequestration. Also, there is a need for active collaboration of government, industry, and farmers to mitigate and adapt for a more resilient agricultural sector [15, 16].

### 2.3.3 Climate-Smart Agricultural and Technological Interventions

Precision farming indicates a transformation of agriculture that provides data-centric recommendations to decision-making, contributing to better field-level management [17]. Tools and technologies, such as drones, equipment guided by global positioning systems, and smart sensors, collect and analyze real-time data to provide insights on crop status, soil health, and resource needs precisely. By enabling targeted solutions with precision inputs such as variable rate application of watering, fertilizing agents, nutrient supplements, and pest control agents, these methods contribute to resource efficiency, increased yield, and alleviation of harmful environmental consequences. Thus, real-time monitoring and tracking of crop health leads to smart decision-making and the use of personalized farming strategies to improve productivity and meet sustainability goals based on their needs.

Data analytics and science, with their complex algorithms and predictive models, need to be part of agricultural practices to improve them. The system, integrated with data analytics, increases crop productivity while reducing resource waste by predicting fertilizer application based on specific soil nutrient status and crop demands. Through personalized assistance, farmers can decide based on the rich data analytics to yield a quality product and market it. Precise water management infrastructure, like drip irrigation systems, enhances water application. Given the increasing global water scarcity, CSA practices support cautious water usage aimed at inducing crop growth. By refining efficient water usage, farmers can achieve sustainable approaches to both irrigation and fertilizer resources, making their farming operations more resilient.

### 2.3.4 Integration of Indigenous Knowledge

Technological modernizations in CSA are more effective when they combine technical knowledge with farmers' regional and localized experience, based on weather conditions. In this regard, the value of indigenous knowledge is especially important, as it offers practical perceptions and time-tested practices that complement modern techniques. The priority on native knowledge confirms that CSA approaches are both context-specific and culturally relevant to improve their acceptance and long-lasting impact. Practices like local seed varieties, intercropping, and traditional

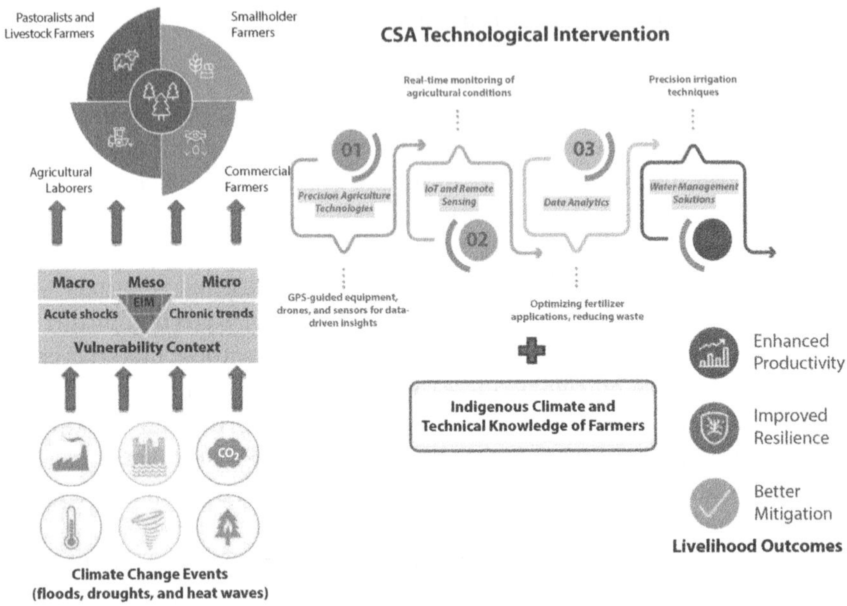


**Figure 2.3** The CSA framework with technology and indigenous knowledge integration.

crop rotation techniques improve soil health and reduce the probability of pest infections. To cope with climate variability, indigenous communities also practice water management techniques, including rainwater harvesting and small-scale irrigation systems [18]. Additionally, methods that support ecosystem health and biodiversity, such as agroforestry and the preservation of sacred groves, are integral parts of Indian culture. Climate-smart agricultural solutions are likely to be inclusive and socially acceptable when Indigenous communities are involved in decision-making processes. This collaborative approach leads to more sustainable and profitable outcomes when local farmers have a sense of ownership. As illustrated in Figure 2.3, the CSA framework integrates both the advanced technologies and indigenous knowledge to improve sustainability.

### 2.3.5 Livelihood Outcomes

The interventions lead to the following significant outcomes:

- **Improved Food Security:** The CSA practices, including precision farming, drought-tolerant crop varieties, and optimized irrigation techniques, increase crop productivity.

This increase in productivity is directly proportional to food availability and accessibility, thus promoting food security.

- **Operational Efficiency:** By integrating modern technologies like remote sensing, geographic information systems, and automated machinery, the efficiency of the farming operations is enhanced. It results in precise monitoring and management, simplifying planting, irrigation, and harvesting. It also reduces input costs and environmental impacts, leading to cost-effective resource management.
- **Resilient Farming Systems:** The deployment of CSA procedures such as agroforestry, soil conservation, and advanced water management leads to a sustainable farming system. Moreover, it mitigates risks associated with extreme weather conditions such as drought and flooding. By using intelligent strategies like rotational cropping, regenerating soil health, and managing water usage, farmers can increase the resilience of their pastures. Such procedures enable farmers to better resist and recover from adversarial environmental conditions, resulting in more reliable agricultural production with increasing climate variability.
- **Livelihood Protection:** The CSA initiatives enhance resilience to climate variability by protecting farmers' livelihoods against the adverse effects of changing climatic conditions. These approaches ensure continuous agricultural productivity and promote economic stability to mitigate the socio-economic risks of rural communities. With CSA, farmers can handle environmental uncertainties to maintain long-term socio-economic development within their regions.
- **Sustainable Development with Reduced GHG Emissions:** The integration of CSA practices, such as minimal soil tillage, regulated fertilizer application, and integrated pest control, reduces GHG emissions from farming activities. By limiting tillage, the physical and chemical characteristics of the soil are increased with greater carbon sequestration, which retains more organic substances in the soil. The optimal usage of fertilizers leads to minimized nitrogen inputs, thereby minimizing $N_2O$ emissions. Integrated pest management reduces chemical pesticides, which improves soil health and maintains ecosystem balance. Also, it reduces

$CH_4$ emissions as well, for mitigating the impact on climate variability and promoting sustainability.

- **Alignment with Global Climate Targets:** The implementation of sustainable agricultural practices directly correlates with the global climate objectives posted in the Paris Agreement. Around 80% of national climate commitments focus more on agriculture to mitigate, adapt, and ensure ecological balance with progress. With this alignment, sustainable goals can be achieved with food security and livelihood support for the rural and suburban communities.

## 2.4 Integration of Artificial Intelligence and the Internet of Things in Precision Agriculture

The combination of smart technologies and connected devices is transforming farming techniques, particularly in areas vulnerable to climate change. Hence, integrating these technologies into precision agriculture improves real-time surveillance and control of agricultural practices. Precision agriculture uses modern farming methods such as equipment, drones, and sensors enabled by global positioning systems to acquire data on soil quality, crop health, and environmental metrics [19]. This wealth of data helps farmers to make strategic decisions concerning resource distribution, including water and fertilizers, leading to better crop yields while minimizing waste. For example, systems that support decision-making can interpret live data and provide practical advice, allowing farmers to promptly respond to weather changes, such as droughts or heavy rains, by modifying their operations. Additionally, IoT devices allow for continuous monitoring and automated irrigation control, ensuring crops receive water at the right time to improve their health and productivity.

### 2.4.1 Wireless Sensor Networks for Soil Monitoring

Wireless sensor networks are formed to collect environmental data from farms and fields to track and observe crop vitality and soil nutrient status. With this real-time data, farmers can be assisted in making accurate decisions aligned with the exact environmental status of their own field [20]. The IoT-based platforms, device-to-cloud data analysis, remote farming access, and real-time device connectivity are developing improved data management across various domains. On attaining preset threshold limits, the sensor readings trigger alerts and notifications to farmers and other entities. Low-cost soil sensors are available to measure temperature data, nutrient levels, moisture levels, and pH levels of the cultivated soil. The best-suited crops for cultivation are predicted by analyzing the raw soil parameters and weather conditions using the K-nearest neighbors algorithm. The farmers can decide on crop rotations most appropriate for their particular soil conditions using this machine learning technology, which classifies soil data by comparing it with historical datasets. The algorithm provides the speedy processing of voluminous datasets and assists with timely crop selection decisions. A continuous crop monitoring system enables continuous crop monitoring based on preset criteria [21, 22]. The system continuously tracks crop growth in phases, climate variables, and vulnerabilities, including pest and disease conditions, using real-time data from wireless sensor networks (WSNs) and smart connected devices. It provides the farmer with accurate information that could assist with proactive management throughout growth.

### 2.4.2 Near-Surface Camera Network for Monitoring within the Climate-Smart Agricultural Framework

An existing work enhances agricultural monitoring, specifically for CSA-related activities, by combining remote sensing technologies with a near-surface camera (NSCam) network [18]. By positioning high-resolution cameras over agricultural areas, as shown in Figure 2.4, the NSCam Network captures images in red-green-blue (RGB) scale at an hourly resolution. The green chromatic coordinate index is then used to analyze the imagery to determine the vegetation status. To facilitate time-series and spatial analysis, the acquired data are paired with enhanced vegetation index 2 (EVI2) satellite measurements. Energy-efficient power solutions, such as solar panels, and real-time uploads in Fourth Generation (4G) technology, facilitate continuous operations. This method offers greater temporal

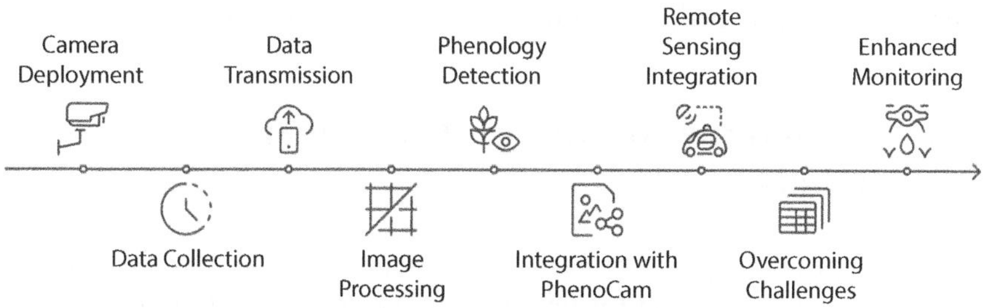


**Figure 2.4** Enhancing agricultural monitoring with NSCam network.

resolution and ground-level precision while overcoming the drawbacks of conventional satellite remote sensing, such as infrequent revisit cycles and vulnerability to cloud cover. However, geographic coverage is limited by the fixed installation of NSCams, and there are logistical and technological challenges in maintaining the cameras and coordinating the data with satellite imagery [23]. Despite these drawbacks, this approach, however, presents several potential implications for long-term phenological research, anomaly identification, and real-time crop care that can be progressively applied to more robust and adaptive farming.

### 2.4.3 Integrating Unmanned Aerial Vehicles and Artificial Intelligence for Precision Agriculture

Next-generation farming is transformed by the combination of unmanned aerial vehicles (UAVs) algorithms, which address the complex challenges of present-day crop management. Using UAVs equipped with state-of-the-art sensors, this approach captures high-resolution pictures of agricultural areas, which are subsequently processed using a classification method that makes use of deep learning algorithms. The approach efficiently detects weeds while allowing for the detection and monitoring of a variety of crops, including rice varieties. When compared with traditional methods, such as Naïve Bayes and Decision trees, this framework outperforms rice seedling classification and weed identification with better classification accuracy and faster convergence when the Adam optimizer is used, particularly [24]. The decentralized learning paradigm, such as federated learning, allows efficient distributed model training with privacy preservation, as raw data is stored locally and only model updates are shared, overcoming privacy and security challenges in data analytics. This decentralized approach improves accuracy with near-real-time data to achieve collaborative learning across scalable agricultural landscapes. However, there exist challenges with infrastructure management, data heterogeneity, and communication limitations in some areas. Also, some issues enforce fairness and trustworthiness among network participants.

## 2.5 The Internet-of-Things-Based Smart Farming Robots

The concept of using IoT-based smart agricultural robots in greenhouses integrates ecological monitoring and autonomous navigation to optimize crop growth [25]. These agricultural robots are a part of the networking participants in WSN that collect temperature, humidity, and soil moisture data. Local processing of these sensor readings leads to trend analysis and prediction using machine learning algorithms, with which the farmers arrive at decisions without accessing the cloud to identify areas of inefficient growth. The robots are integrated with Proportional-integral-derivative controllers for accurate navigation over rough terrain. These robots can collect extensive and voluminous data without disturbing the ongoing farming processes. With their high-precision environmental sensing nature, the robots can make immediate adjustments in response to slight variations in climate and soil. This approach provides a range of advantages, such as optimal resource usage through precise watering and fertilizing at the exact time of requirement and in the precise location. This precision increases crop yields and reduces labor costs through automated machinery. However, the system involves investment costs and necessitates specialized technical expertise in the operational efficiency and technical maintenance of the subsystems. There are risks associated with high-end technologies, such as operational interruptions due to connectivity issues or device malfunctions, especially in harsh greenhouse conditions.

## 2.6 Linking Weather and Climate Information Services with Climate-Smart Agriculture

Recent work explores how CSA practices are integrated with weather and climate information services (WCIS) to enhance agricultural water management and resiliency, especially in the global south [27]. It supports several CSA practices by rainwater harvesting, growing a variety of crops, making compost, and using integrated pest management techniques. To reinforce these CSA practices, WCIS provides weather forecasts on seasonal, monthly, and daily scales. As per the claim, seasonal forecasts are advantageous for making strategic decisions, such as crop and its variety selection, land distribution, and irrigation planning. The daily forecasts guide in making operational actions such as scheduling irrigation, managing pests, and field surveillance. The temperature and rainfall data are shared through extension services, which support taking timely actions. It includes carbon-smart using composting, water-smart by improved varieties, upgraded irrigation, weather-smart by taking crop insurance, and nitrogen-smart using organic fertilizer, and crop diversification. These recommendations are closely timed with rainfall and temperature patterns to increase their efficiency. The research highlights the need to customize WCIS to associate with different agro-ecological conditions and operational challenges at various scales. This process also includes refining water circulation protocols at the watershed level, supplementing crop sequencing strategies for single agricultural units, and managing *in situ* rainwater collection systems. The analysis identifies continuing difficulties such as insufficient digital proficiency among farmers, lack of infrastructure, and the need for localized WCIS solutions, even as technological evolutions continue. To overcome these problems, there should be direct investments for digital revolutions, the adoption of inclusive policies that address gender inequalities, and the formation of participatory governance models. Remote sensing and AI are also used to improve data delivery and usability. Scaling WCIS access and transforming it with CSA practices significantly improves sustainability by providing policymakers and practitioners in climate-vulnerable areas with a plan of action.

### 2.6.1 Machine Learning for Weather Forecasts

Machine learning techniques are important for agricultural weather forecasting because they can analyze extensive datasets and find recurring trends and patterns that affect crop production. These techniques produce precise predictions suited to particular agricultural requirements by utilizing real-time sensor inputs, satellite imaging, and historical weather data,

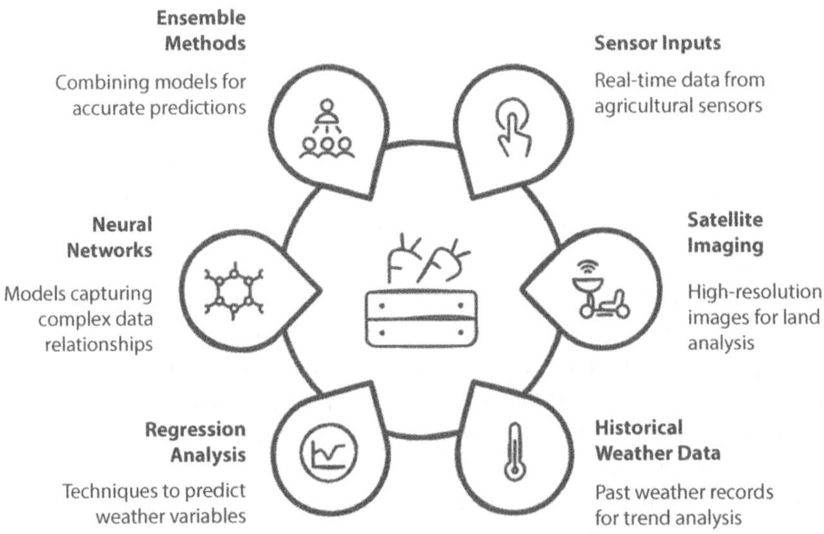


**Figure 2.5** Machine learning in CSA.

as illustrated in Figure 2.5. Machine learning can improve the accuracy of forecasts for temperature, precipitation, and humidity by utilizing strategies such as regression analysis, neural networks, and ensemble methods. Farmers can enhance crop productivity and resource utilization by using these predictive capabilities to optimize planting dates, irrigation techniques, and pest management plans. Additionally, when new data is processed, machine learning models can adjust over time, continuously improving their accuracy [26]. However, the quality and granularity of these algorithms determine their effectiveness.

## 2.7 Advanced Water Management Techniques

Advanced water management techniques, particularly precision irrigation, significantly impact water usage efficiency and crop productivity within the framework of CSA. Techniques such as drip irrigation and the deployment of soil moisture sensors allow farmers to apply water more judiciously, directly correlating with improved crop yields and reduced water wastage. Research indicates that implementing precision irrigation can save 20–50% of water while increasing crop productivity by up to 30%. Optimizing resource usage is more important to face climate variability on a positive note. Inconsistent meteorological trends cause the downfall in the reliability of traditional agricultural practices. With the implementation of precision irrigation and tools, farmers increase their adaptive capacity to environmental variations and contribute to sustainability objectives.

### 2.7.1 Optimizing Water Consumption through Artificial Intelligence-Based Irrigation Management

This technique optimizes irrigation with smart sensors and actuators by using a predictive algorithm that uses water more efficiently and minimizes excessive percolation deep into the soil. It works by processing empirical data to assess water retention capacity in the soil and then recommends a probabilistic confidence interval in scheduling irrigation. The layered architecture comprises three layers where three different machine learning models are trained: a multi-layered perceptron neural network, a support vector machine, and a K-neighbors classifier. These learning algorithms learn from multimodal data such as soil characteristics, environmental parameters, and meteorological forecasts for the next three days. The novel platform predicts watering requirements with almost 99% accuracy. Its efficiency was validated through Monte Carlo simulations, which demonstrated reductions in energy consumption of up to 57% and water usage by 27%. This approach is beneficial in precise forecasting, adaptability across various crop types, and enhanced water management efficiency. However, there are also some challenges, including initial setup costs, the necessity for farmers to acquire specialized skills, and data privacy issues while sharing information. By mitigating the challenges associated with resource limitations, this prototype also enhances sustainable farming practices [27].

### 2.7.2 Internet of Things-Based Smart Irrigation Systems

The deployment of an IoT-based smart irrigation system includes an array of basic components designed to ensure more efficient water usage in agricultural practices. This architecture intrinsically integrates cloud-based infrastructure with WSNs to enable continuous data acquisition and periodical data analytics in real-time. A centralized cloud infrastructure collects data from sensors that record environmental parameters such as temperature, humidity, and soil moisture for analysis. The cloud-native infrastructure uses big data analytics to generate real-time observations, supporting farmers to reach data-driven decisions about irrigation and resource planning. The technique also combines renewable energy sources, such as solar power, to develop sustainability and decrease dependency on conventional energy sources. By integrating meteorological prediction services, as shown in Figure 2.6, the dynamic adjustment of irrigation strategies can be made in response to anticipated precipitation, enabling optimal water resource management during variable climatic patterns. This approach increases agricultural productivity, minimizes functional expenditures, improves environmental sustainability, and enhances effective water utilization [28]. Through precise irrigation and real-time monitoring, water wastage can be minimized and provide more efficient resource utilization. Although there exist some challenges, like substantial investment costs associated with acquiring and setting up sensors and subscribing to cloud-based resources. Additionally, the compatibility in integrating various subsystems requires specialized training for farmers. Without suitable analytics, the volume of information produced can be excessive, making data governance another difficulty. The effectiveness of these solutions may be severely restricted in rural locations by challenges with connectivity, privacy, and data security implications. Although IoT-based smart irrigation systems feature a viable method to support sustainable agriculture, their effective deployment requires a complete evaluation of the positive and negative aspects.

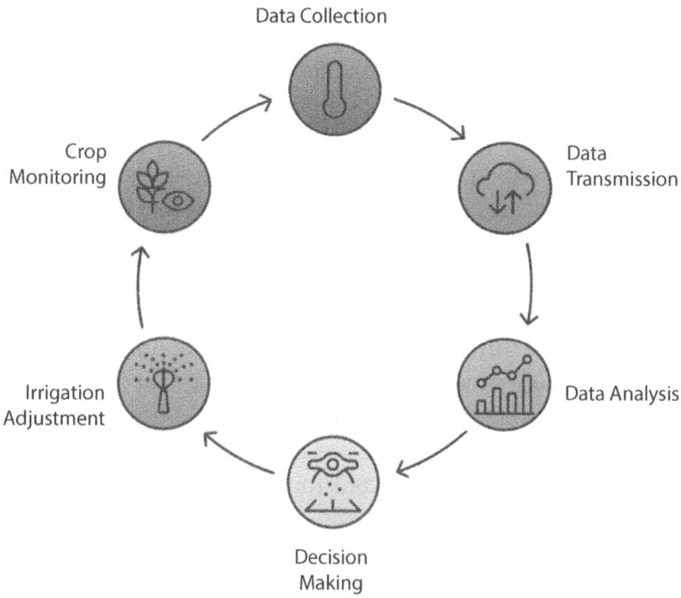


**Figure 2.6** An IoT-based smart irrigation cycle.

## 2.8 Contribution of Agroforestry Practices and Renewable Energy

Agroforestry practices and renewable energy solutions are needed for improving biodiversity and soil health, thereby promoting sustainability in farming systems affected by climate change. The agroforestry method combines trees and shrubs with grazing, improving soil quality by composing organic matter and essential nutrients, providing habitats for diverse fauna, and sequestering atmospheric $CO_2$. Such an integrated approach improves biodiversity and immunity against pests and disease challenges. Also, renewable energy solutions like solar-based irrigation reduce dependence on fossil fuels, maintaining ecological balance. By following these approaches, the farmers can design and develop more sustainable, adaptable, and economical agricultural models. These novel models ensure long-term productivity and adaptability to evolving climate variations.

### 2.8.1 Precision Land Management for Agroforestry

There are two main ways to evaluate the appropriateness of land for climate-smart agroforestry (CSAF) implementation: (i) the analytical hierarchy process and (ii) geographic information systems. The land under consideration is classified into optimal, moderate, and unsuitable zones. It is based on biophysical and socioeconomic features like soil characteristics, altitude, rainfall, and access to infrastructure. This classification process optimizes land management while mitigating climate change by guiding informed decisions in CSAF implementation. The practices of agroforestry enhance soil fertility, control soil erosion, support biodiversity, and sequester carbon consistently with adaptive climate methodologies. There is proof that CSAF improves food security and restores damaged lands in Bugesera and Rulindo. Also, there are disadvantages, including a lack of awareness among farmers, restricted market accessibility of agroforestry

**Table 2.2** Summary of the identified key inferences.

| Research focus | Inferences |
|---|---|
| AI and IoT in Precision Agriculture | Real-time monitoring, Resource optimization, Data-driven decisions |
| Advanced Water Management | Precision irrigation, Automated irrigation systems Variable Rate Sprinklers, Evapotranspiration controllers |
| Agroforestry and Biodiversity | Nutrient cycling, Carbon sequestration, Nitrogen fixation, Microclimate regulation |
| Renewable Energy in Agriculture | Biofuels, Agrivoltaics, Greenhouse heating, Soil sterilization, Floating solar farms, Solar dryers |
| Indigenous Knowledge Integration | Polycultures, Rotational farming, Rainwater harvesting, Terracing, Floodplain farming, Biochar utilization, Composting, Mulching |
| Impact of CSA Technologies | CSA practices align with global climate goals, ensuring sustainability, resilience, and improved food security. |
| Challenges and Adoption Barriers | High establishment costs, technical expertise gaps, and limited digital literacy hinder CSA adoption, especially in developing regions. |

goods, and delayed returns on investment. The expansion of agroforestry practices is limited by some constraints, such as conflicts over land use, the complexity of integrating tree species into current farming practices, and high startup costs. In certain areas, steep slopes and poor soils make agroforestry unsuitable. The identification of kinds of literature is presented in the following Table 2.2.

## 2.9 Conclusion

This chapter explores the adoption of modern technologies such as AI and IoT for precision agriculture and irrigation alongside traditional farming practices. It is evident from prior literature that this integrated approach improves resource efficiency, reduces GHG emissions, and increases resilience to climate variability. In regions that are more vulnerable to climate variability, it is important to integrate real-time monitoring with live resource management. The implementation of satellite-based remote sensing, data-driven learning algorithms, AI applications, and IoT-enabled precision agriculture platforms enhances agricultural output, optimizes irrigation schedules, and maintains ecosystem stability. The CSA practices, like agroforestry and renewable energy sources, increase sustainability by promoting biodiversity, enhancing soil fertility, and supporting ecological resilience. To scale these technologies globally, it is necessary to overcome socio-economic challenges and implement policy frameworks that involve farmers for sustainable agricultural development. However, socioeconomic, cultural, and technological challenges restrict the transformative potential of CSA. To contribute to global sustainability goals and ensure food security and resilience in agricultural systems, future research should concentrate on the following areas: developing scalable and accessible CSA technologies for smallholder farmers; working with policymakers to create supportive frameworks; applying indigenous knowledge for region-specific solutions; expanding the use of AI and machine learning for predictive analytics; creating comprehensive sustainability metrics; and investing in education and training programs to empower farmers and stakeholders.

**AI disclosure statement:** The author confirms that no artificial intelligence (AI)-generated content or AI-assisted writing tools were used in the preparation of this chapter (including summaries and the abstract). All text, analysis, and conclusions are the result of the author's original work.

## References

1. Ma, W. and Rahut, D.B., Climate-smart agriculture: adoption, impacts, and implications for sustainable development. *Mitig. Adapt. Strateg. Glob. Chang.*, 32, 457, 2024.
2. Gemtou, M., Kakkavou, K., Anastasiou, E., Fountas, S., Pedersen, S.M., Isakhanyan, G., Erekalo, K.T., Pazos-Vidal, S., Farmers' Transition to Climate-Smart Agriculture: A Systematic Review of the Decision-Making Factors Affecting Adoption. *Sustainability*, 32, 234, 2024.
3. Pagano, A., Croce, D., Tinnirello, I., Vitale, G., A survey on LoRa for smart agriculture: Current trends and future perspectives. *IEEE Internet Things J.*, 10, 3664, 2023.
4. Adli, H.K., Remli, M.A., Wong, K.N., Ismail, N.A., González-Briones, A., Corchado, J.M., Mohamad, M.S., Recent advancements and challenges of An IoT application in smart agriculture: A review. *Sensors*, 23, 2453, 2023.
5. Zhao, J., Liu, D., Huang, R., A review of climate-smart agriculture: Recent advancements, challenges, and future directions. *Sustainability*, 24, 543, 2023.
6. Alzubi, A.A. and Galyna, K., Artificial Intelligence and Internet of Things for Sustainable Farming and Smart Agriculture. *IEEE*, 11, 78686, 2023.
7. Kong, J., Fan, X., Jin, X., Su, T., Bai, Y., Ma, H., Zuo, M., BMAE-Net: A Data-Driven Weather Prediction Network for Smart Agriculture. *Agronomy*, 32, 458, 2023.
8. Zerssa, G.W., Feyssa, D.H., Kim, D., Eichler-Löbermann, B., Challenges of Smallholder Farming in Ethiopia and Opportunities by Adopting Climate-Smart Agriculture. *Agriculture*, 11, 19, 2021.
9. Qazi, S., Khawaja, B.A., Farooq, Q.U., IoT-Equipped and AI-Enabled Next Generation Smart Agriculture: A Critical Review, Curr. Chall. Fut. Trends. *IEEE Access*, 10, 21219, 2022.
10. Jin, X., Zhang, J., Kong, J., Su, T., Bai, Y., A Reversible Automatic Selection Normalization Deep Network for Predicting in the Smart Agriculture System. *Agronomy*, 12, 675, 2022.
11. Shaikh, F.K., Karim, S., Zeadally, S., Nebhen, J., Recent trends in Internet-of-Things-enabled sensor technologies for smart agriculture. *IEEE Internet Things J.*, 9, 23583, 2022.
12. Yang, X., Shu, L., Chen, J., Ferrag, M.A., Wu, J., Nurellari, E., Huang, K., A survey on smart agriculture: Development modes, technologies, and security and privacy challenges. *IEEE/CAA J. Autom. Sin.*, 8, 273, 2021.
13. Kumar, P., Gupta, G.P., Tripathi, R., PEFL: Deep Privacy-Encoding-Based Federated Learning Framework for Smart Agriculture. *IEEE Micro*, 42, 33, 2022.

14. Sakapaji, S.C. and Puthenkalam, J.J., Harnessing AI for Climate-Resilient Agriculture: Opportunities and Challenges. *Eur. J. Theor. Appl. Sci.*, 45, 678, 2023.
15. Tsakiridis, N.L., Diamantopoulos, T.G., Symeonidis, A.L., Theocharis, I.B., Iossifides, A.C., Chatzimisios, P., Pratos, G., Kouvas, D., Versatile Internet of Things for Agriculture: An eXplainable AI Approach. *J. Artif. Intell.*, 584, 180, 2024.
16. Ouafiq, E.M., Saadane, R., Chehri, A., Data management and integration of low power consumption embedded devices IoT for transforming smart agriculture into actionable knowledge. *Agriculture*, 34, 236, 2022.
17. Senapaty, M.K., Ray, A., Padhy, N., IoT-Enabled Soil Nutrient Analysis and Crop Recommendation Model for Precision Agriculture. *Comput.*, 12, 61, 2023.
18. Yu, L., Du, Z., Li, X., Zhao, Q., Wu, H., Weise, D., Yuan, X., Yang, Y., Cai, W., Song, W., Wang, P., Zhao, Z., Long, Y., Zhang, Y., Peng, J., Xin, X., Xu, F., Shen, M., Wang, H., Jiao, Y., Li, T., Sun, Z., Zhao, Y., Fang, M., Peng, D., Wu, C., Li, S., Shen, X., Ma, K., Lin, G., Luo, Y., Near surface camera informed agricultural land monitoring for climate smart agriculture. *Clim. Smart Agric.*, 54, 678, 2024.
19. Mabhaudhi, T., Dirwai, T.L., Taguta, C., Senzanje, A., Abera, W., Govid, A., Dossou-Yovo, E.R., Aynekulu, E., Petrova Chimonyo, V.G., Linking weather and climate information services to Climate-Smart Agriculture (CSA) practices. *Clim. Serv.*, 34, 213, 2025.
20. Rosero-Montalvo, P.D., Gordillo-Gordillo, C.A., Hernández, W., Smart Farming Robot for Detecting Environmental Conditions in a Greenhouse. *IEEE*, 11, 57843, 2023.
21. Pavithra, A., Kalpana, D.G., Vigneswaran, D.T., Deep learning-based automated disease detection and classification model for precision agriculture. *Soft Comput.*, 31, 12, 2023.
22. Mukhamediev, R.I., Yakunin, K., Aubakirov, M., Assanov, I., Kuchin, Y.I., Symagulov, A., Levashenko, V.G., Zaitseva, E.N., Sokolov, D., Amirgaliyev, Y.N., Coverage path planning optimization of heterogeneous UAVs group for precision agriculture. *IEEE*, 11, 5789, 2023.
23. Shaikh, T.A., Rasool, T., Lone, F.R., Towards leveraging the role of machine learning and artificial intelligence in precision agriculture and smart farming. *Comput. Electron. Agric.*, 198, 107119, 2022.
24. Dong, Y., Werling, B.P., Cao, Z., Li, G., Implementation of an in-field IoT system for precision irrigation management. *Front. Water*, 13, 245, 2024.
25. Seyar, M.H. and Ahamed, T., Development of an IoT-Based Precision Irrigation System for Tomato Production from Indoor Seedling Germination to Outdoor Field Production. *Appl. Sci.*, 42, 321, 2023.

26. Makuvaro, V., Chitata, T., Tanyanyiwa, E., Zirebwa, S., Challenges and Opportunities in Communicating Weather and Climate Information to Rural Farming Communities in Central Zimbabwe. *Weather Clim. Soc.*, 2023.
27. Bungău, S.G., Behl, T., Aleya, L., Bourgeade, P., Aloui-Sossé, B., Purza, A.L., Abid, A., Samuel, A.D., Expatiating the impact of anthropogenic aspects and climatic factors on long-term soil monitoring and management. *Environ. Sci. Pollut. Res.*, 28, 30528, 2021.
28. Luo, Y. and Pu, L., UAV Remotely-Powered Underground IoT for Soil Monitoring. *IEEE Trans. Industr. Inform.*, 20, 972, 2024.

# 3

# Grokking Deep Reinforcement Learning for Climate-Smart Agriculture

**N. Mythili[1]*, V. Saranya[2], P. Manjula[3] and Raffaele Mascella[4]**

*[1]Department of Physics, Sri Venkateshwaraa College of Engineering and Technology, Ariyur, Puducherry, India*
*[2]Department of Computer Science and Engineering, Sri Venkateshwaraa College of Engineering and Technology, Ariyur, Puducherry, India*
*[3]Department of Health Information Management and Technology, College of Applied Medical Sciences, King Faizal University, Alahsa, Saudi Arabia*
*[4]Department of Communication Sciences, University of Teramo, Teramo, Italy*

***Abstract***
Predicting crop yield and plant pathology, weed detection, and soil/water management have been the cardinal research areas. Abrupt shifts in climatic patterns that adversely impact crop management can be subdued by adopting resilient strategies with recent technologies. Deep learning (DL) and reinforcement learning (RL) could fix issues related to yield prediction and suggestions based on global/regional climatic change. The downside of DL is that it is inadequate to generate a mapping between raw data and the production value. Deep reinforcement learning (DRL)-based algorithms will prevail over DL; they provide the finest platform for adopting climate change strategies. This chapter discusses the diverse modeling approaches based on DL and RL algorithms used in climate-smart agriculture (CSA), which predict and boost productivity. To ensure crop and livestock management, pest and weed control, various levels of monitoring, and modeling approaches based on DRL, which have an impact on the food supply chain and agro-economies, were elaborated. Various resilient methods/processes were discussed globally in light of new crop varieties, crop diversification strategies to ensure water and soil quality, and soilless cultivation.

***Keywords***: Deep reinforcement learning, deep learning, climate-smart agriculture, modeling approaches, monitoring systems

**Corresponding author*: mythuwinmile@gmail.com; ORCID: https://orcid.org/0000-0002-3854-6995

Anitha Velu, Prasanth Aruchamy, Raghu Ramamoorthy, Rajesh Kumar Dhanaraj and Seifedine Kadry (eds.) Climate Smart Agriculture, (41–60) © 2026 Scrivener Publishing LLC

## 3.1 Introduction

Revolutionary changes in agricultural practices across different eras, with various objectives, gave rise to terms such as the Green Revolution (Grain production), White Revolution (Milk Productivity), Silver Revolution (Egg Productivity), etc. This era is labeled the 'Digital Revolution' in agriculture with the objective of achieving a high yield in a diversified environment and climatic conditions, enhancing food security to meet global demands. The demand for food and agro products is predicted to rise to 70% in 2050 [1]. Doubling the population, tripling food demand, reducing land area, aggravated climatic change and its effects, imbalance between micro- and macronutrient levels in soil, eutrophication effect on water bodies and their related salinity problem, impact of pesticides on biomagnification are some of the intensified issues in modern agriculture. Major shifts in weather and climatic patterns as a consequence of an increase in average temperature and altered seasonal patterns in all layers and rainfall patterns. This shift in climatic pattern has a major impact on the biomes of species and their related diseases, which in turn leads to poor production, less yield during demand due to environmental calamities, loss of agricultural land in flash floods, landslides, drought, crisis, among others. The foremost victim of this scenario is poor, marginalized, and small-scale farmers, for whom the primary source of income is agriculture. Climate change and its subsequent effects led to the migration of agronomists to suburban and urban areas. This makes skillful farming professionals go off track and reduces the labor force in agriculture.

To address this issue, connecting sustainable agricultural practices with developing technology is the demand of the hour. Climate-smart agriculture is a greater initiative in balancing the environmental and economic outlook, facilitating innovation and technology to adapt to the transition of new climatic patterns with less human intervention. Climate-smart agriculture is an agrarian system that helps to include flexible agricultural practices and to ensure national and global food safety with reduced greenhouse gas (GHG) emissions. This chapter briefly discusses the policies, outlines, and impacts of CSA in the digital revolution, highlighting its major goal in sustainable farming. Various modeling approaches and monitoring systems used in precise farming based on approaches based on deep learning (DL) and reinforcement learning (RL), and their impact on policy and decision-making predictions are analyzed. Diverse algorithms based on deep reinforcement learning (DRL), which are used in all forms of agricultural practices, starting from field mapping, seed sowing, and supply chain logistics to ensure a transformative change in farming, are explored. Adaptation strategies and the future of revolution in CSA with resilient practices and mitigation measures in accordance with climatic variations are the purview of this present chapter.

## 3.2 Panoramic Perspective of Climate-Smart Agriculture

### 3.2.1 Climate-Smart Agricultural Policy

Climate change and its related issues and impacts were signaled in the early nineteenth century. But then it hit global concern in the First World Climate Conference held in 1979. This conference sparked global discussions and actions. It established an advisory group to standardize the Intergovernmental Panel on Climate Change in 1988, in connection with the World Meteorological Organization. Various sub-agencies and non-governmental organizations engaged in creating global awareness on GHG emissions and monitoring. In the early 1990s, the major shortcoming was the difference between developed and developing countries' targets in controlling GHG emissions. According to reports, the developed countries' targets for controlling emissions are minimal. Hence, to stabilize the situation, the "Kyoto Protocol" was introduced, which includes policies and methods that encourage developed countries to reduce their GHG emissions by 5.2% by 2012.

In 2008, the European economy faced massive backlash and decelerated actions regarding climate change. In this circumstance, developing countries like India, Brazil, and China are making steady progress in taking actions to reduce GHG emissions. The Copenhagen Accord came into action in 2009 and states that, in correlation to industrial levels, the global temperature rise must be within 2 °C and developing countries must report their action taken in this regard every two years [2]. Thus, based on this accord, both developing and developed countries must set baselines to reduce GHG emissions based on their frameworks. After 2020, climate-based policy documents have deepened their roots, the impact of climate change on the agricultural sector has been taken into account, and various mitigation and adaptation strategies have been framed and processed.

### 3.2.2 Outline of Climate-Smart Agriculture

The concept of CSA was first proposed in the Food and Agriculture Organization report in 2009. This concept is widespread globally because it addresses solutions to numerous crucial issues, such as food scarcity during natural disasters, economic issues in the agro-industry, ensuring an eco-friendly environment, barriers to accessing technology, and infrastructural investments. Various resilient methods/processes were tested globally under the roof of CSA, such as new crop varieties that can withstand heat and salinity, crop diversification strategies to ensure water and soil quality [3].

The major goal of CSA is to develop solutions to regional or domestic issues and a low carbon footprint. Location/region-specific research is the need of the hour to understand the feasibility, socio-economic aspects, adaptation barriers, etc. At present, developing countries like India, China, and Malaysia have already implemented appropriate interventions to resilient technologies based on CSA, which requires minimal manpower and risk. Climate-smart agriculture intends to achieve a "Triple Win" by the Food and Agriculture Organization by ensuring the following:

i) Sustainable productivity by adapting to climate change;
ii) Altering and developing the technologies that are resilient to climate variations; and
iii) Reducing GHG emissions.

### 3.2.3 Climate-Smart Agriculture as Sustainable Farming

The agriculture sector is a key area affected by climate change, not only because of its vulnerability to climatic effects but also due to food security. The climatic flip-flops have a direct influence on the backlashes faced by the agricultural sector and directly impact the nation's gross domestic product. Climate-smart agriculture does not suggest new principles/processes; rather, it integrates precision and resilient strategies into sustainable agriculture practices, financial investments, and policies. Certain policies of CSA include conservation/protection of natural resources, uplift of local livelihoods, and escalating income and productivity. The major actuation of CSA is to develop a sustainable environment and agricultural governance through proper balancing of resources and technologies.

Information and communication technology-based investments will strengthen the farmer's learning and sharing ideas to support decisions in the farming process. Eventually, the utilization of the ideas and implementation will be estimated by efficiency in integrating climatic response with agricultural processes in real time. If this CSA practice is considered seriously, then an environmentally sustainable agricultural practice with socio-economic benefits and trade-offs will be the outcome after 20 years.

## 3.3 Big Data

Soil, crop, field mapping, Soc, nutrient technology-based farming are inescapable in the current era due to dramatic changes in climate and soil nutrients. Implementing concepts of the Internet of Things (IoT), machine learning (ML), DRL, robotics, and drones in farming can ease monitoring soil, crops, and real-time environmental conditions for precision farming. Impacts of various modeling approaches, such as big data analysis, ML, DL, and RL algorithms based on CSA, are discussed below.

### 3.3.1 Data Collection

In any monitoring system, the initial phase is data collection. To address appropriate techniques for data collection in smart or precision farming, field sensors, geographic sensors, global positioning systems, and high-resolution remote sensing play a key role in validating spatial and zonal variations. The major big data applications in CSA are soil quality and weather monitoring, irrigation motorization, market demand and understanding, seed and polyculture analysis, pest and fertilizer monitoring, crop monitoring, weed control, etc. It is shown in Figure 3.1. Involving big data analysis in smart farming is a recent concept, although real-time field sensors will bridge the gap between derived data integrated with large data sources available globally.

Sensor nodes are connected to microcontrollers, wireless devices include Wi-Fi, Zigbee, long-range wide-area networks, and narrowband-IoT, where long-range wide-area networks provide long-range connectivity, and narrowband-IoT, Wi-Fi provides wider bandwidth [4]. Numerous data related to farming and weather conditions, which will be condition-supportive of understanding and building the data structure, have limited access and reporting [5]. A range of valid information on bygone, predictions, and real-time conditions has been processed with big data techniques for CSA. Big data analysis is the collective process of edge computing, data processing and transmission layers, cloud computing, and sequential decision-making (SDM), which are discussed below.

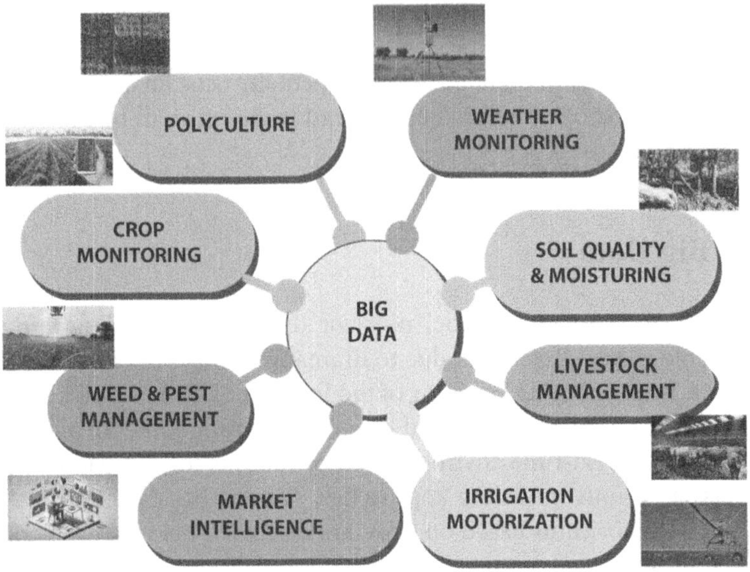


**Figure 3.1** Big data analysis and applications in CSA.

### 3.3.2 Edge Computing

In data collection, processing, and decision-making in smart agriculture, edge computing is evolving as a promising model. Edge computing will help process real-time data close to the data source, increasing responsiveness. Edge computing is classified into preprocessing and postprocessing. Preprocessing state is analyzing each sensor and detecting the incomplete data from the datasets. The role of the post-processing state is to compress a huge volume of data to a minimal dataset. Thus, post-processing state helps minimize the bandwidth required to transfer data to the cloud, which will be the next step in smart farming. Edge computing will help reduce the volumes of data collected from real fields to minimal data by sampling methods.

### 3.3.3 Data Transmission Layer

The collected data from data fusion will be technically transferred to the cloud layer by various data mining algorithms and artificial intelligence (AI) protocols. The IoT system enables this data transfer. Multiple data from various field sensors will be integrated using data fusion techniques. This technique improves data reliability by fusing multi-sensor data through a special algorithm to obtain fused or unfused datasets. Different stages of data fusion include signal, pixel, feature, and decision levels [6]. In general, the data fusion process starts with reading data given by various sensors, processing the data, extracting features, and fusing [7]. The major outlook of data fusion with IoT is single and multi-hub data. In a single-hub fusion system, the sensor data interacts only with the fusion system. Whereas, in a multi-hub, various sensor data were fused with a multi-fusion system. The suggested method in data fusion is DRL since it can learn continuously. The DRL approach will furnish reliable decision-making regarding irrigation and nutrient management, crop and weed monitoring, field mapping etc.

### 3.3.4 Cloud Computing and Sequential Decision-Making

Data analysis based on environmental conditions is an important and key area in precision farming. This will be executed using DRL to suggest smart decisions. Sequential decision-making is essential in DRL since it has achieved superhuman or remarkable performance in crop management. "Stopping rule" is an important term in SDM since it decides when to stop and start taking observations, which helps to minimize the losses in observation costs. The stopping rule decides when to start/stop taking observations, and this decision-making will help reduce loss and maximize gain. The detailed flowchart from data collection to SDM is given in Figure 3.2.

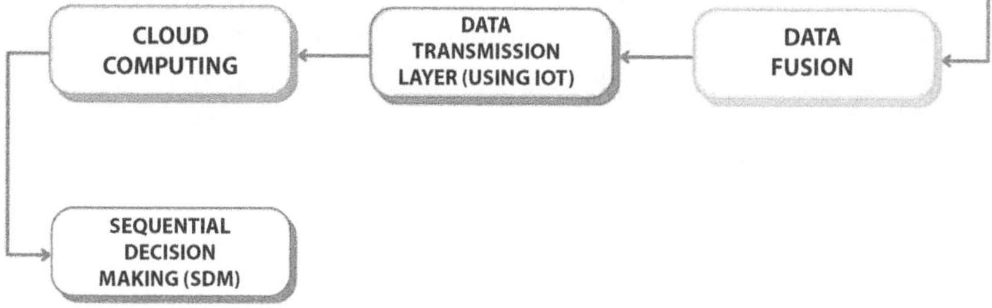


**Figure 3.2** Flow chart of data collection to SDM in big data analysis.

## 3.4 Machine Learning

Machine learning learns from existing datasets without precise programming. The datasets are classified as train and test data, which assist in training and predicting, respectively. Machine learning has various algorithms, from traditional learning to deep learning. Among them are random forest, support vector machines, and decision trees, which have been used in ML-based CSA. Machine learning has two broad categories: supervised and unsupervised learning algorithms. Labeled data is the prime factor in supervised learning, which includes algorithms related to classification and regression. Similarly, unlabeled data for correlations is used in unsupervised learning, which has clustering and dimensionality-related algorithms.

A special category of ML is semi-supervised learning, a technique for comparing a few labelled data points with huge, unlabeled data. In semi-supervised learning, among various algorithms most employed in CSA is K-means clustering since it helps in object detection by improving the image quality, ensuring pest and weed management. The major drawback of ML in agriculture is that tuning algorithms based on requirements within pipelines, streamlining algorithm models for specific tasks, etc. As of now, ML algorithms have been opted to maximize productivity, fertilizer usage, and rates, to evaluate the soil properties, and so on [8, 9]. But traditional ML techniques require manual model selection and feature selection, which may lead to less accurate predictions.

Deep learning requires large datasets, which are difficult to achieve in the agriculture sector. Even if it is attained, data reliability and accuracy may be frail depending on various environmental factors; it will be overdue by AutoML. It selects the algorithm automatically based on the task by rationalizing the model development. "No free lunch" theorem emphasizes that no algorithm will excel in all problems; this is taken into account for predications related to CSA. The foremost thing in using ML is choosing the appropriate algorithms. In considering all these, the AutoML technique may suit CSA, although it can access advanced analytical techniques with standardizing procedures and frameworks. The AutoML tool results in accurate diagnostics of weed, soil fertility, and plant diseases, critically evaluating output based on climatic conditions for irrigation management, and analyzing the crop selection.

## 3.5 Deep Reinforcement Learning

Deep reinforcement learning is a fusion of DL and RL in finding out the challenges in SDM. This, in turn, makes the machines mimic humans when solving problems based on sequential decisions. Deep reinforcement learning develops the ideas for various decision-making using networks.

### 3.5.1 Deep Learning

Large datasets and improved computational performance make DL an effective technique in precision farming. Deep learning specializes in grid-based data; it has a multi-layer data structure, which can be in the format of images or data. It should have an input and output layer, which are visible layers, and a hidden layer. This technique requires fewer human interventions; it automatically extracts the features from raw data based on feature learning, and it also forms a higher-level hierarchy with lower-level features is one of its unique features. Diverse architecture and algorithms like convolution neural networks (CNN), recurrent neural networks (RNN), long-short term memory as supervised learning, and self-organizing maps, autoencoders, restricted Boltzmann machine, and generative adversarial networks (GAN) as unsupervised learning.

#### *3.5.1.1 Convolution Neural Networks*

Convolutional neural networks recognize the related features automatically without human intervention. They also help in large-scale network implementation with multiple building blocks. For image processing and associated extractions, like plant diseases and yields to estimate the quality of fruits and vegetables, CNN is the best model because it is self-trained and has superior computational efficiency. The CNN model for cocoa bean classification and wheat maize yield predictions was reported; the inference from this is that sensing index and patch/area size, number of deep convolution layers play an important role in predictions [10, 11]. The quality grading of fruits, such as mangoes, bananas, and apples, is an important aspect in smart farming.

Developing a technical solution for the standard grading system will help the sellers and buyers understand fruit quality. It also helps farmers check the export quality of products to maintain their reputations in global markets. Non-destructive testing with imaging technologies (CNN) embedded with microcontrollers helps attain a technical solution for quality grading. The CNN architecture in smart farming consists of four layers, and the schematic view of the CNN architecture is given in Figure 3.3.

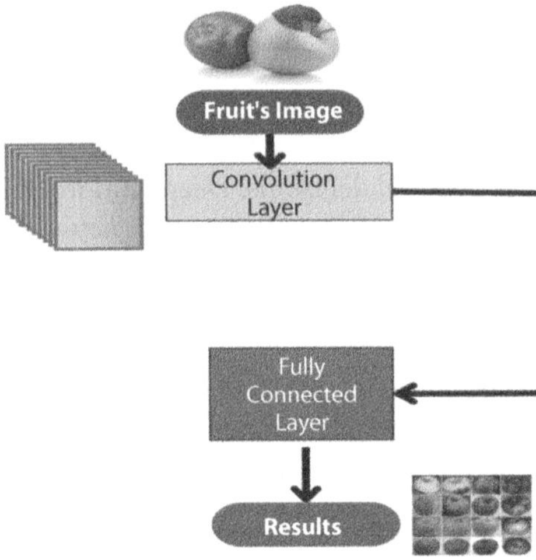


**Figure 3.3** The schematic view of the CNN model in image processing.

*Convolution layer*: Once the input image is received, this layer extracts image features in color, intensity, shape, and texture, among others. This leads to the creation of a feature map.

*Rectified linear unit layer*: Non-linearity in the image is elevated with the help of the rectified linear unit layer, which in turn resolves gradient issues. This layer helps the model acquire a more complex representation of the image. It transforms input value into positive numbers, and the mathematical representation with the help of an activation function [12] is given in Equation 3.1, where '0' is the least value and '$x$' represents a positive number.

$$f(x)_{ReLu} = (0, x) \tag{3.1}$$

*Pooling layer*: This layer helps minimize the obtained information without losing major features in simple terms. This layer assists in creating a sub-sampling of the created feature map by using various pooling methods, such as tree pooling, average pooling, min/max pooling, etc. The pooling layer is a major step in CNN because it should not fail to include important features, as it affects the output.

*Fully connected layer*: This layer will cumulate the predictions of the pooling layer into a single vector and generate the final predictions of images, such as classification. Inside this layer, neurons of the pooling layer are connected with every neuron; hence, it is fully connected. The output of this layer stands as the CNN output. Convolutional neural networks follow this general architecture to find their application in CSA by image processing.

### 3.5.1.2 *Recurrent Neural Networks*

The RNN algorithms in CSA use sequential data because it delivers precious information about crop classification based on their characteristics, climatic adaptability, growing seasons, and soil requirements. The RNN algorithm is trained based on training data, which helps to optimize the model parameters. Training data can be obtained from preprocessed data in which the collected data is classified as training and testing sets. The RNN architecture consists of an input layer, a hidden layer, and an output layer, where each layer has its own neurons and activation functions. The initialization of weights and biases (W&B), where 'W' represents the strength of the input data, and 'B' is the added information to proceed effectively, is the first step of the RNN. Input, hidden, and output layers are added after this initialization. The constructed model is trained on training datasets, and W&B are adjusted or optimized to decrease the loss function. The RNN model's performance is evaluated using testing data. This model helps identify healthy crops, classify crops, and detect weeds, etc. [13–15].

### 3.5.1.3 *Generative Adversarial Networks*

Data collection based on agricultural parameters, crops, and yields is multitudinous due to an unstructured environment, crop and weed diversity, plant disease variants, different irrigation practices, etc. To overcome these difficulties, robust technical support with large datasets and a wide range of imaging conditions should be preferable. Traditional image augmentation techniques need numerous image modifications/transformations. Image augmentation based on DL is processed using GAN, a promising way to learn more variations. The two-layer architecture of GAN includes the generator and the discriminator. The generator model helps create complex data samples. In other words, these techniques synthetically generate data that will be similar to raw data. The discriminator differentiates between raw data and the created dataset by functioning as a binary classifier. The workflow diagram of the GAN model is given in Figure 3.4. Steadily, it learns to differentiate between created and raw data. The GAN model helps in image/data augmentation of shape, style, crop color, weed, etc. This enhances weed detection [16, 17], which in turn helps productive yield by minimizing human intervention, both physically and technically.

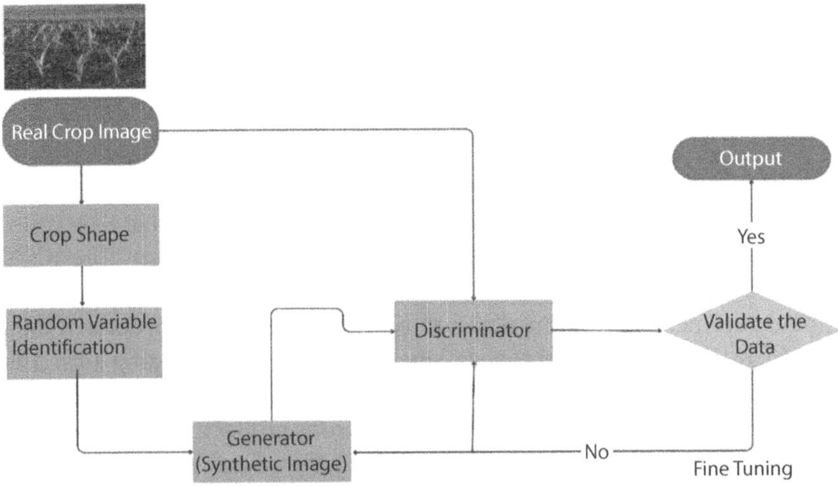


**Figure 3.4** Workflow diagram of the GAN model.

### 3.5.2 Reinforcement Learning

Reinforcement learning is like a trial-and-error method that attains knowledge from environmental interactions. The main aim of RL is to learn the strategy that will maximize the reward. Reinforcement learning is based on making independent decisions like autonomous agents with respect to the environment without human involvement and shares the relation between agents, goals, and the functioning environment. This approach is formulated in terms of the Markov decision process (MDP), and the schematic view of the MDP process is shown in Figure 3.5.

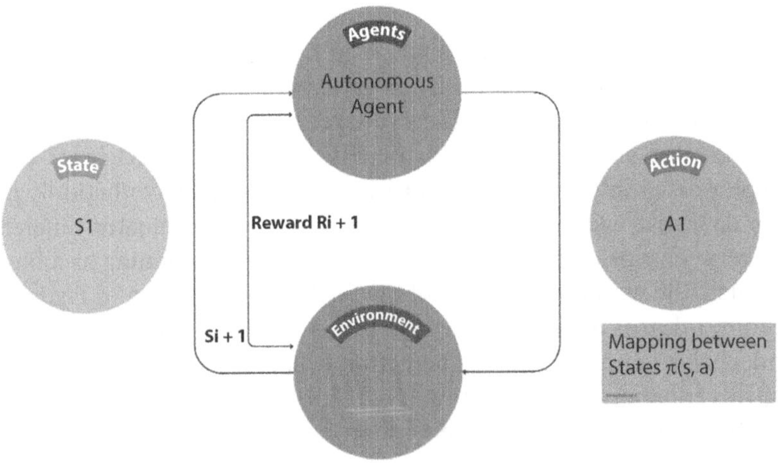


**Figure 3.5** Schematic view of the MDP process.

**Table 3.1** Elements of MDP and its representation.

| S. no. | MDP elements | Mathematical representation |
|---|---|---|
| 1. | Sets of states | $S$ |
| 2. | Sets of Actions | $A$ |
| 3. | Rewards | $R$ |
| 4. | Each state is represented by | $S_I$ |
| 5. | Mapping between states | $\pi\,(s,a)$ |
| 6. | So, the function depends on | $\sum_a \pi\,(s,a) = 1$ |

The MDP elements are listed in Table 3.1, where 'state' spaces mean the set of data driven from the environment, whereas 'action' spaces refer to the set of all actions taken by the agents, and the 'reward' function is the feedback received by the agents for the actions done within a state. The RL algorithm will help in field mapping using an unmanned aerial vehicle (UAV) [18] to understand the correlation between past production methods vs new approaches [19].

## 3.6 Various Monitoring Systems Using Deep Reinforcement Learning in Climate-Smart Agriculture

### 3.6.1 Crop Monitoring, Field Mapping Using Deep Reinforcement Learning

Deep reinforcement learning mainly uses either pixel or object-based detection. In these, hyperspectral imaging plays a vital role because it aims to analyze variations in the ground with respect to time. It also helps recognize pathogens, nutrient supply, and identify crops. This imaging is very precise, helping to identify levels of anthocyanin or chemical compositions in plants using least-square support vector and extreme learning machine. Crop monitoring, field mapping, and yield analysis-based data collection are crucial in predictions because they involve various inputs in the form of pixels, objects, and data. Hyper-spectral imaging is seizing this task in precision farming by its accuracy in spectral response.

### 3.6.2 Seed Sowing and Water Management-Based Deep Reinforcement Learning

Relating to policy-making and decisions, Proximal Policy Optimization (PPO) is the best tool in DRL due to its coherence and stability. This method focuses on updating the policies using the agents to make decisions. An optimal policy network is attained by training the agent to small policy updates for reliability. This also checks that the current policy is not too different from the old policy to maintain clarity. At the initial stage, the data is gathered using old policies associated with the environment. Then, optimizing policy decisions is achieved by increasing the objective function. Additionally, the advantage function is an important factor in policy-making algorithms since it explains the improvement in action with respect to the old policy. This also exhibits the quality of action taken and permissible change ratio in the existing policy. Equation 3.2 is the relation between yield (*Y*) and field parameters (*SD, W, D, P*).

$$Y = f(SD, W, D, P) \tag{3.2}$$

where *Y*-yield of the field, *SD*-seed sowing density (SD), *W*-amount of water, *D*-sowing depth, and *P*-policy. Integrated with agricultural bots, PPO is a major tool in CSA since the decisions are based on climatic conditions. The real-time implementation of Agrorobotix integrated with PPO in agriculture suggests an increasing trend in yield of up to 16.3% and a decreasing trend of water usage up to 21.7% [20].

### 3.6.3 Pest, Weed Detection/Management Using Deep Reinforcement Learning

While recapitulating DRL algorithms based on weed detection, nearly 50% of the work opted are based on CNN algorithms with the efficiency of 99.29% in predicting crops and weeds [21]. Various CNN architectures were tried in the reported literature, such as AlexNet [22], Inception V3 [23], Xception, and DenseNet [24]. Convolutional neural networks integrated with UAV and laser technology/smart sprayer will reduce manpower in weed management. A laser weeding system based on $CO_2$, diode, and fiber lasers will predict the weeds in the field and take action based on the algorithm. Similarly, a smart sprayer will target the weed area and ensure precise spraying of pesticides at the prescribed concentrations. For past identification, the transfer learning model (AlexNet) is promising as it requires a small dataset, and reliable predictions were reported [25].

### 3.6.4 Fleet Management and Logistics Using Deep Reinforcement Learning

Due to the maximum utilization of machines in smart farming, it is essential to map the relationship between owning and operating vehicles. Fleet management helps schedule the routing, tracking, and supervising the maintenance of vehicles, resource distribution, planning the vehicle utilization based on performance and efficiency, and weather conditions in the field. Fleet management requires sequential decision-making since it learns from a trial-and-error strategy. Similarly, policy-based and value-based algorithms will speed up the learning process, and thus, Soft Actor-Critic algorithms gear up the interface of fleet management. The Soft Actor-Critic algorithm is based on a continuous process with the possibility of checking maximum combinations to obtain optimal results [26]. Similarly, a voice-driven fleet management system and 3D tracker acquire real-time information about field and vehicle routing, etc.

The extended version of fleet management has a greater impact on supply chain logistics, and its efficient functioning is the mainstay of agroeconomics. Supply chain logistics is a complex function because it involves production, processing, storage, transport, retail distribution, and exporting. The deep deterministic policy gradient algorithm in DRL is an actor-critic approach that helps learn even in offline training, storing past experiences, and handling multi-dimensional states. Due to its learning capability, the deep deterministic policy gradient algorithm will help interface with supply chain systems to bridge the gap between producers and consumers, enhance the sustainability, customer services, and resilience during calamities. The reports suggest that using the above algorithm in supply chain logistics increases the inventory turnover, quality metrics, and prompt delivery [27]. The PPO algorithms can also be engaged in synchronization issues related to the supply chain.

### 3.6.5 Livestock Management Using Deep Reinforcement Learning

Livestock and farming are inextricable; they both contribute to agroeconomics. Livestock-based farming using IoT is considered precise and smart farming, and IoT-based livestock management helps in optimized grazing patterns and breeding practices. It helps monitor the feed, track livestock, and detect diseases in early stages to strengthen the stock population. Global positioning systems, radio frequency identification-based tags, smart collars, UAVs, and drones can be involved to make this technically transparent.

In livestock management, implementing DRL algorithms flawlessly, IoT is a predominant aspect since IoT leads to the proliferation of DRL architectures with compatible standards. An arrowhead framework is a service-oriented architecture that bridges the gap between developing hardware and network platforms. Similarly, digital twin-based DRL helps stimulate real-world situations and predict the physical counterpart. This helps to plan and optimize the form process based on livestock management.

## 3.7 Adaptation and Alleviation Strategies Under a Climate Change Scenario

Various adaptation strategies based on CSA include crop selection and planting by season, pest and weed detection and extraction, fleet and water management with respect to requirement, supply chain, and livestock maintenance based on demand and usage. Technically alleviating the impacts of climatic variations on agriculture is a pressing priority, as it directly impacts countries' gross domestic product and food chains. Various machines/robots enabled with IoT exist in agricultural practices such as seeding/weeding robots, harvesting bots, pest control sprayers, soil testing robots, and palletizing robots for well-ordered logistics, etc., drones and sensors for monitoring and spotting the yields.

Combining sensors, 3D construction methods, and image processing methods will create high-definition, realistic, and virtual 3D models to enhance the monitoring approach. Drones are a major game-changer in precision farming. The agriculture-based drone market is expected to increase globally from $4.17 billion in 2022 to $18.22 billion in 2030. Drones and UAVs facilitated by navigation systems perform multiple functions with a fine accuracy ratio. An AI-based interface with weather sensors and IoT helps ensure proper irrigation management with precise decision-making. Deep learning methods and their implementation in farming have greater potential due to their precision and predictions.

## 3.8 Future Scope of Deep Reinforcement Learning in Climate-Smart Agriculture

Climate-smart agriculture aims for sustainable practice from all layers of agriculture, starting from field mapping to product logistics and supply chain by making precise predictions using AI algorithms that interface with embedded technology. The productivity of AI algorithms depends on data accuracy and quantity; hence, robust infrastructure is required to ensure large-scale data collection and storage. An IoT-based smart farming must be customizable and adaptable to local stakeholders because the need for technology usage varies with respect to regions. Integrating meteorological services like automated weather stations, online weather services, and cloud forecast data with IoT is a prime concern that will enhance service quality and help achieve the CSA objectives. Soilless cultivation might be the future of CSA due to a lack of space, nutrients, and vulnerable climatic conditions. Hydroponics (growing plants using water nutrients), Aquaponics (a hybrid of hydroponics and aquaculture), vertical farming, and substrate farming are a few categories of soilless cultivation that have already started rooting in various regions. Artificial intelligence-driven soilless cultivation, bots, and UAVs will change today's farming into more precise farming.

## 3.9 Conclusion

Climate-smart agriculture emphasizes sustainable agricultural productivity and economy in alignment with recent technologies in farming. Both DL- and RL-based modeling approaches, such as CNN, RNN, GAN, and RL, help in policy and decision-making predictions more precisely. Various algorithms, such as PPO, hyper-spectral imaging, and MDP, help in irrigation management, planting and seed-sowing process, pest and weed management, etc. From field mapping to fleet and supply chain management, deep learning-based algorithms will assist in enriching the implementation of CSA without constraints on regions and climates due to their predictions and precision. Eventually, the utility of the concepts, policies, and performance will be assessed by their response to sustainable agricultural development, resilient practices with mitigation measures in accordance with climatic variations. In case mitigation measures and policies aligned with AI technology in farming are implemented in the fields to the fullest, in 20 years, it can be ensured that a transformative change in economic, environmental, and socio-economic perspectives.

**AI disclosure statement:** The author confirms that no artificial intelligence (AI)-generated content or AI-assisted writing tools were used in the preparation of this chapter (including summaries and the abstract). All text, analysis, and conclusions are the result of the author's original work.

## References


1. World Bank Group, 2020, https://openknowledge.worldbank.org/handle/10986/9122.
2. Joyeeta, G., *A history of international climatechange policy*, vol. 1, p. 636, John Wiley & Sons, WIREs, Climate Change, 2010.
3. Adjaye, J.A. and Yohannis, T., Leveraging climate-smart agriculture to address climate risk in Africa. *ACET*, 14, 521, 2019.
4. Wang, J., Analysis of wireless communication networks under edge computing scenarios. *Wirel. Netw.*, 28, 3665, 2022.
5. Ünal, Z., Smart Farming Becomes Even Smarter with Deep Learning - A Bibliographical Analysis. *IEEE Access*, 8, 105587, 2020.
6. Shi, P., Li, G., Yuan, Y., Kuang, L., Data fusion using improved support degree function in aquaculture wireless sensor networks. *Sens.*, 18, 3851, 2018.
7. Alam, F., Mehmood, R., Katib, I., Albogami, N.N., Albeshri, A., Data fusion and IoT for smart ubiquitous environments: A survey. *IEEE Access*, 5, 9533, 2017.
8. Peng, W. and Karimi Sadaghiani, O., A review on the applications of machine learning and deep learning in agriculture section for the production of crop biomass raw materials. *Energy Sources Part A Recover. Util. Environ. Eff.*, 45, 9178, 2023.
9. Mohamed, S.A., Metwaly, M.M., Metwalli, M.R., AbdelRahman, M.A.E., Badreldin, N., Integrating Active and Passive Remote Sensing Data for Mapping Soil Salinity Using Machine Learning and Feature Selection Approaches in Arid Regions. *Remote Sens.*, 15, 1751, 2023.
10. Yudhi, A., Setya, W.P., Mario, K., Jenq-Shiou, L., Convolutional Neural Network Application in Smart Farming. *Soft Com. Data. Sci.*, 14, 287, 2019.
11. Petteri, N., Nathaniel, N., Tarmo, L., Crop yield prediction with deep convolutional neural networks. *Com. Elect. Agric.*, 163, 104859, 2019.
12. Alzubaidi, L., Zhang, J., Humaidi, A.J., Al-Dujaili, A., Duan, Y., Al-Shamma, O., Santamaría, J., Fadhel, M.A., Al-Amidie, M., Farhan, L., Review of deep learning: Concepts, CNN architectures, challenges, applications, future directions. *J. Big Data*, 8, 53, 2021.
13. Brahim, J., Loubna, R., Noureddine, F., RNN- and CNN-based weed detection for crop improvement: An overview. *Food. Raw. Mater.*, 9, 387, 2021.
14. Emile, N., Dinh Ho, T.M., Nicolas, B., Dominique, C., Laure, H., Deep Recurrent Neural Network for Agricultural Classification Using Multitemporal SAR Sentinel-1 for Camargue. *France Remote Sens.*, 10, 12172, 2018.
15. Loubna, R., Brahim, J., Noureddine, F., Lekbir, A., Belaid, B., Digital transformation metamodel in smart farming: Crop classification prediction based on recurrent neural network. *Food. Raw Mater.*, 13, 1, 2024.

16. Mulham, F., Vincenzo, S., Daniele, N., Domenico, D.B., Shape and Style GAN-Based Multispectral Data Augmentation for Crop/Weed Segmentation in Precision Farming. *Crop Prot.*, 184, 106848, 2024.
17. Mulham, F., Ciro, P., Ibis, P., Alberto, P., Domenico, D.B., Daniele, N., Data Augmentation Using GANs for Crop/Weed Segmentation in Precision Farming. *IEEE Conf. Con. Tech. App.*, 17, 142, 2020.
18. Saeid, P. and Ardakani., A.C., Reinforcement Learning-Enabled UAV Itinerary Planning for Remote Sensing Applications in Smart Farming. *Telecom.*, 2, 255, 2021.
19. Gabriela, E.N.E., A Reinforcement Learning Approach for Smart Farming. *Database Sys. J.*, 8, 103, 2019.
20. Abdullah, A., Muhammad, T., Najma, I., Enhancing Precision of Crop Farming Towards Smart cities: An Application of Artificial Intelligence. *Sustainability*, 16, 355, 2024.
21. Hu, W.Z., Wane, S.O., Zhu, J.K., Li, D.S., Zhang, Q., Bie, X.T., *et al.*, Review of deep learning-based weed identification in crop fields. *Int. J. Agric. Biol. Eng.*, 16, 1–10, 2023.
22. Krizhevsky, A., Sutskever, I., Hinton, G.E., ImageNet classification with deep convolutional neural networks. *Advances NIPS*, vol. 32, p. 165, 2012.
23. Szegedy, C., Ioffe, S., Vanhoucke, V., Alemi, A., Inception-ResNet and the Impact of Residual Connections on Learning. ArXiv.,160207261, 2016.
24. Huang, G., Liu, Z., Van Der Maaten, L., Weinberger, K.Q., Densely connected convolutional networks. *IEEE Con. Com. Vis. Patt. Rec.*, 23, 2261, 2017.
25. Wang, D., Deng, L., Ni, J., Gao, J., Zhu, H., Han, Z., Recognition pest by image-based transfer learning. *J. Sci. Food Agric.*, 99, 4524, 2019.
26. Lisang, L., Jionghui, C., Youyuan, Z., Jiayu, C., Jingrun, L., Dongwei, H., Unmanned Ground Vehicle Path Planning Based on Improved DRL Algorithm. *Electronics*, 13, 2479, 2024.
27. Murugeshwari, B., Mohanapriya, M.P., Brindha Merin, J., Akila, R., A Deep Reinforcement Learning Approach for Optimizing Inventory Management in the Agri-Food Supply Chain. *J. Electr. Systems*, 20, 2238, 2024.

4

# Understand Cutting-Edge Reinforcement Learning Algorithms for Controlled Environment Agriculture

**Udayakumar K.[1]*, Revathi M.[2], Sharmila L.[3] and Muhammad Rukunuddin Ghalib[4]**

*[1]Department of Computer Science and Design, Vel Tech Rangarajan Dr. Sagunthala R&D Institute of Science and Technology, Chennai, India*
*[2]Department of Computer Science and Engineering, Bharath Institute of Higher Education and Research, Chennai, India*
*[3]Department of Computer Science and Engineering, Vel Tech Rangarajan Dr. Saguthala R&D Institute of Science and Technology, Chennai, India*
*[4]Department of Cyber Security, De Montfort University, Dubai, United Arab Emirates*


## *Abstract*

An economically viable agricultural system is the essential cornerstone of a growing economy. There is a need to employ cutting-edge technology to enhance crop quality, scrutinize crop diseases, and forecast yields. The main thrust here lies in the realm of reinforcement learning (RL) to shape agriculture, because one might realize that RL autonomous systems are built to maximize resource efficiency, streamline agricultural practices, and enhance crop yield. This chapter explored cutting-edge RL algorithms for controlled environment agriculture (CEA). Integrating RL into CEA may bring about a revolutionary transformation. It may allow for the development of an autonomous decision-making system that can adapt to dynamic environmental conditions, optimize the use of resources, and maximize crop output and quality. This chapter is targeted towards researchers, engineers, and professionals working in the domains of artificial intelligence, agricultural technology, and environmental control systems. This study explores how cutting-edge RL algorithms enhance CEA systems, addressing current and future challenges and prospects. It aims to inspire further research and advancements in applying RL to agricultural practices.


**Corresponding author:* drudayakumar@veltech.edu.in

Anitha Velu, Prasanth Aruchamy, Raghu Ramamoorthy, Rajesh Kumar Dhanaraj and Seifedine Kadry (eds.) Climate Smart Agriculture, (61–84) © 2026 Scrivener Publishing LLC

***Keywords*:** Controlled environment agriculture, digital twin, reinforcement learning, sustainable agriculture

## 4.1 Introduction

Globally, agriculture is the primary buttress of a country's economy and livelihood. A truth that cannot be hidden is that agriculture contributes a significant portion of the country's gross domestic product (GDP). Precisely 17% of India's GDP is contributed by the agricultural sector. Moreover, it is the country's largest source of employment. The potential initiatives have been consistently made in agriculture policy to ensure organic farming, sustainability, self-reliance, and a friendly ecosystem. Regrettably, traditional farming practices have been affected because of diverse climates, global warming, soil depletion, and chemical fertilizers. Traditional methods have undergone many challenges. Specifically, low productivity is due to a lack of enhanced seed variants. In addition, it has a significant impact on infrastructure, people migration, and market access.

Today's broad technological advancements offer more solutions to mitigate the challenges of traditional agriculture. The term "agritech," recently coined, encompasses various technologies and devices. That includes drones, artificial intelligence (AI), soil sensing, remote sensing, weather monitoring, and machine learning (ML) [1]. These immense innovations transform traditional agricultural methods into highly sustainable and proficient methods. It supports farmers in managing farms in various ways, such as indoor farming, wastewater treatment, precision farming, and farm automation. Even though traditional farming practices attain rich sustainability and eco-friendliness, several challenges remain, such as unpredictable weather, low production, a lack of resources and infrastructure, and migration to urban areas.

Now, farmers have embarked on modernized farming practices through adopting technologies, as illustrated in Figure 4.1. In the revolution of modern agriculture, controlled environment agriculture (CEA) has become an inevitable form of agriculture. Climate variations influence traditional agriculture, whereas in CEA, farming is in a controlled, closed environment. Controlled environment agriculture can be adopted for many reasons, such as reducing water usage, less land, increasing crop protection, consistent yield, and reducing pesticide use [2]. It is also ideal for urban farming.

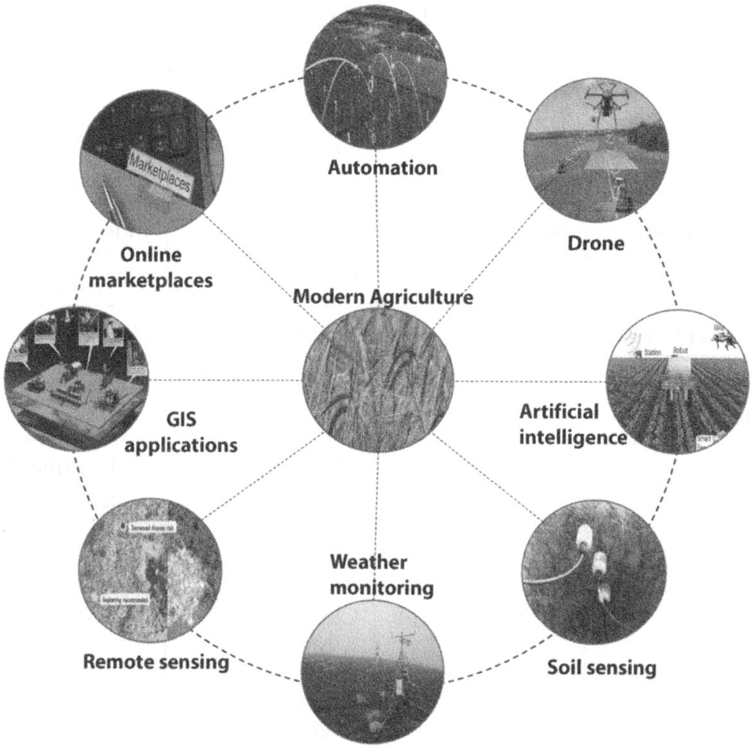


**Figure 4.1** Modern farming practices.

### 4.1.1 The Role of Controlled Environment Agriculture

Controlled environment agriculture integrates Internet of Things (IoT) devices, actuators, controllers, sensors, and AI-driven systems to create an optimal ecosystem for cultivation. Eventually, it provides year-round production, biosecurity, and accurate control over light, temperature, humidity, carbon dioxide, and nutrients. Controlled environment agriculture can curtail the costs associated with traditional agriculture. The idea of CEA opens the possibility of more localized food production in urban areas. Table 4.1 summarizes key aspects of traditional agriculture and CEA.

The challenges of food security can be addressed through innovations in vertical farming. Artificial intelligence-driven analytics and automation will likely increase productivity and enhance the role of CEA globally. As technology grows, CEA will become more accessible, efficient, and sustainable.

**Table 4.1** A summary of traditional agriculture and CEA.

| Aspect | Traditional open-field agriculture | Controlled environment agriculture | |
|---|---|---|---|
| | | Greenhouse farming | Vertical farming |
| **Environment** | Outdoor | Indoor | Indoor |
| **Climate** | Dependent | Independent | Independent |
| **Water-efficiency** | Low | Moderate | High |
| **Pest and Disease** | Higher exposure | Low exposure | Low exposure |
| **Lighting** | Sunlight | Mostly sunlight, artificial lighting if necessary | Only artificial lighting |
| **Land use** | Open field | Closed horizontal | Closed vertical |
| **Crop Yield** | Seasonal yields | Year-round yields | Year-round yields |
| **Energy requirement** | Low | Moderate | High |
| **Technology** | Less usage | Moderate usage | High reliance |
| **Resource efficiency** | Low | Moderate | High |
| **Scalability** | Limited scalability | Limited scalability | High scalability |
| **Setup cost** | Lower | Lower | Higher |
| **Operational Cost** | High | Moderate | Less |

### 4.1.2 Significance of Automation in Controlled Environment Agriculture

Automation is a transformative key in CEA by enabling efficient, precise, and scalable agricultural practices. The modern innovation of automation crucially helps in addressing food security challenges and agricultural needs. The significant aspects of automation are presented in Table 4.2.

**Table 4.2** The significance of automation in CEA.

| Aspect | Technology | Use case |
|---|---|---|
| Improved efficiency | IoT and control algorithms | Moisture levels are used to control water flow in a drip irrigation system. |
| Stability in Growth Conditions | IoT, data analytics, and advanced control algorithms | Smart heating, ventilation, and air conditioning systems adjust settings based on weather forecasts and plant requirements. |
| Early Detection | ML and deep learning (DL) systems | Artificial intelligence-driven platforms analyze images to detect pest infestations or nutrient deficiencies. |
| Supply Chain Management | Radio-frequency identification | Radio-frequency identification systems track harvested produce and automate reordering of supplies. |
| Higher Productivity | AI | Artificial intelligence platforms select traits for faster growth in controlled environments. |

## 4.2 The Notion of Reinforcement Learning in Controlled Environment Agriculture

Controlled environment agriculture highly relies on real-time monitoring systems that utilize large sensor data. Many sensors in closed fields generate more data related to crop health and the environment. Modern technologies like ML, genetic algorithms, and big data analytics are playing crucial roles in employing CEA. For instance, a farmer can plan their marketing choices based on the forecast quality and quantity of yield. Recently, ML has been widely used to optimize CEA. It helps make data-driven decisions and self-controlled automation. Machine learning algorithms precisely deliver core data about crop health, growing patterns, and optimal environmental conditions. The key applications of ML in CEA are climate control, energy optimization, yield forecast, automation, health monitoring, and nutrient management. Using training scenarios, the ML model is instructed on how to do a specific task [3–5]. The system performance is evaluated based on how well it can do the same task in circumstances not present during the training period. In ML, overfitting is a prevalent issue. A phenomenon occurs when, following training, a model performs well in training circumstances but poorly in situations it has not encountered before. Adaptation and continuous learning make ML even more efficient and sustainable in CEA. In addition, it will pave the way for smart and robust agricultural systems.

In parallel with ML, deep learning (DL) algorithms potentially enhance CEA by offering advanced competencies for data analysis, forecasting, and automation. The neural network structure of DL ensures precise insights, which are supportive in managing crops within controlled environments. In addition, image-based crop-monitoring systems draw more attention owing to the convolutional neural networks in DL [6, 7]. To accurately predict agricultural yield, time series analysis can be carried out with the assistance of long short-term memory (LSTM), gated recurrent units, and recurrent neural networks. The DL-powered robotic system helps in automated harvesting and weeding. In summary, the DL algorithm is a pioneer in offering precise data-driven solutions to address vast real-time problems related to CEA.

Recently, reinforcement learning (RL) has become more popular among researchers due to its learning ability in dynamical systems. It is an important paradigm in controlling dynamical systems that are uncertain and unknown. In particular, RL works well for making choices about crop management. Some key areas where RL is making a difference in CEA are illustrated in Figure 4.2.

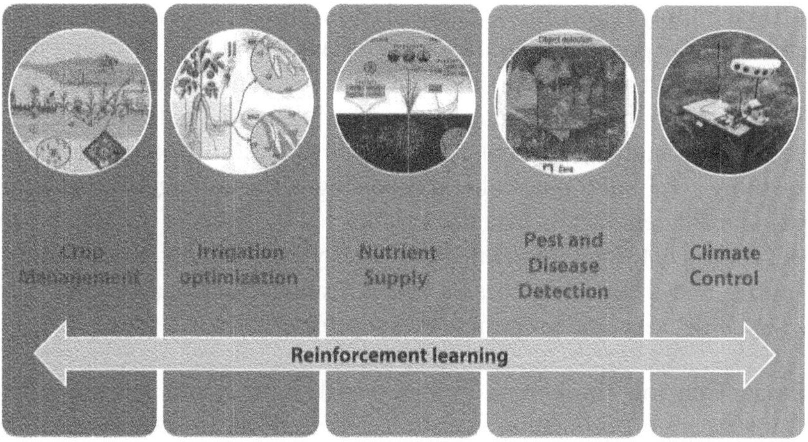


**Figure 4.2** Applications of RL in CEA.

Unlike the ML model, in RL, decisions are made sequentially by interacting with an environment repeatedly [8–11]. The subsequent section addresses the challenges, algorithms, and components of RL. This section focuses on discussing recent literature on RL in CEA.

**Crop Management:** Decision support systems are a core part of crop management. Reinforcement learning could fix some of the problems people have with crop management [9, 12, 13]. It is an interactive, action-oriented, and real-world tool for evaluating a set of unclear crop operations. Reinforcement learning can support decision support systems by learning the optimal policy for harvesting, pruning, and planting. Its ability to adapt to different types of soil and weather helps to get the best quality and yield.

**Irrigation Optimization:** Since irrigation utilizes a lot of water in agriculture, it should be handled differently. For food production, irrigation scheduling is crucial since it guarantees a harvest every time and lowers the possibility of losses due to insufficient water. In agriculture, DL models are already being applied well and are receiving significant attention. Authors trained a deep Q-network (DQN) [14] to determine when to water plants. In their work, they have employed two LSTM networks. The amount of soil water is estimated using one network. The yield is predicted using the other one under the weather. Both LSTM and convolutional neural networks are employed in Q-table estimation during agent training. The trained model reduced water use by 20–30% and boosted output by 11% when compared to the threshold fixed irrigation approach.

**Nutrient Supply:** The authors proposed an RL algorithm to help plants grow in the best way possible, considering important factors [15]. By using RL in farming, it might be possible to fine-tune the climate and the way resources are distributed so that crops produce the most while using the fewest resources possible. The study talks about a framework that uses RL to improve plant growth by combining physical and modeling parts. An "agent" that learns to handle different factors that affect plants is at the heart of the system. The exact control of the environment is possible by using temperature, humidity, moisture, and chemical sensors, as well as heating, cooling, adjustable lighting, water delivery, and nutrient delivery systems.

**Climate Control:** The authors have proposed an integrated system for optimum temperature regulation in a simulated greenhouse environment [16]. Multi-agent RL is used in incorporating intricate environmental models, communication-capable agents, and hyperparameter optimization. The greenhouse environment model provides agents with an enhanced and more comprehensive state representation by incorporating relevant climate factors, plant health metrics, and resource levels for each zone. The reward function has been modified to enhance both individual zone performance and overall greenhouse objectives. This encompasses objectives such as maintaining a stable environment, optimizing resource use, or attaining a balanced yield.

**Pest and Disease Control:** The study [17] presents an adaptive spraying decision system to address the absence of scientific guidance in aerial pesticide application, assisting operators of unmanned aerial vehicles in establishing appropriate spraying parameters in complex environments. The actor-critic neural network-based RL decision model is the core component of the decision system. The critic and actor networks were trained using the field experimental data to adaptively optimize the output of the spraying parameters.

In addition to the uses, certain studies encompass energy efficiency and robotic control, wherein Reinforcement Learning contributes to productive agricultural practices.

## 4.3 Fundamentals of Reinforcement Learning

The ML model adheres to the idea of learning, wherein the trained model provides optimal predictive outcomes for familiar scenarios. However, the model needs to be retrained with reference data to handle new scenarios. Reinforcement learning works on the principle of combining exploitation (knowledge-based learning) and exploration (trial and error). The agent-environment interaction in RL promotes improved learning chances and the application of learned information. The primary function of an agent is to acquire decision-making skills. It engages with the environments to optimize the cumulative reward. The fundamental components of the RL are shown in Figure 4.3. It consists of the agent, environment, observation, action, reward, and policy. This section discusses RL aspects within a scenario of CEA.

**Agent:** A software module that deals with the decision-making process. Regarding the current state, it adjusts environmental parameters like humidity, temperature, irrigation, nutrient supply, light, and $CO_2$ levels. Real-time data of these variables is captured from sensors mounted in vertical/greenhouse farming. These data are fed to the RL agent to make real-time decisions in the form of action. In every training episode, the agent learns to choose optimal action, which creates suitable conditions for plant growth. It controls the actuators based on sensor data that adhere to resource minimization, such as energy, nutrients, and water. For example, an agent might increase the $CO_2$ injection level during photosynthesis.

**Environment:** A place where sensors and actuators are equipped. In our case, an environment is a greenhouse/vertical farming area with installed sensors and actuators. Sensors are deployed to obtain real-time environmental data. Actuators function based on sensor data to control irrigation and nutrient delivery systems, and climate control.

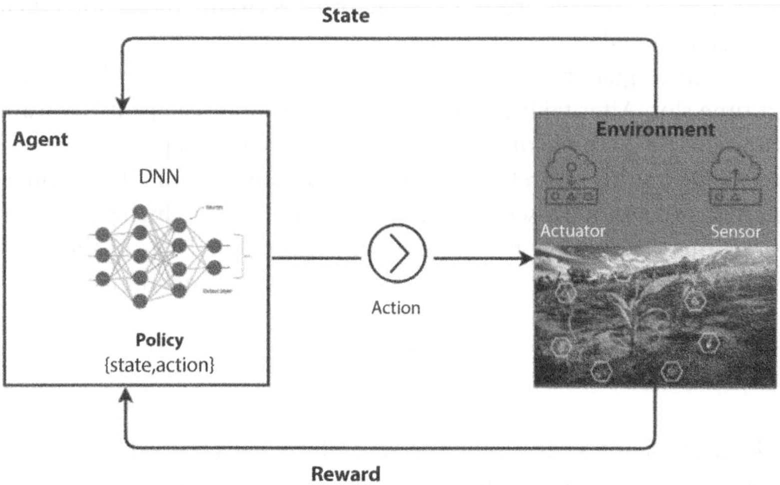


**Figure 4.3** Reinforcement learning in CEA.

**State:** A representation of the current situation in an environment. The state encapsulates all relevant information needed for the agent to decide. The state assists an agent in determining the feasible options for its subsequent action. The agent primarily lacks memory capacity. Consequently, the decision is derived from an analysis of the current condition. At the commencement of each time cycle, an agent discerns the present condition of the environment. A state q in the set $Q$ provides the status information specified in Equation 4.1.

$$\begin{aligned} Q &= \{q \mid q = (d_i, r_i, m_{i,max}, w_i, T_{m,i}, V, S_{all}, H) \\ Q &= \{q \mid q = (Humidity, moisture\ level, Temperature, \\ &\qquad Water\ availability, Crop\ height)\} \end{aligned} \tag{4.1}$$

**Action:** A set of feasible actions that the agent can make based on the available knowledge in a particular state. Actions have consequences and impact the subsequent state of the environment. An agent might control each actuator based on the current state information. It also assesses each action through the defined reward function. The chosen course of action is given in Equation 4.2.

$$\begin{aligned} B &= \{b \mid b = (\mathrm{O}_d, \mathrm{E}_s, N)\} \\ B &= \{b \mid b = (water\ adjustment, nutrient\ supply)\} \end{aligned} \tag{4.2}$$

**Reward:** In RL, the reward is a key factor in verifying the action's quality of an agent. Based on the observed state at time slot t, the policy network is modified to guarantee that the agent makes the optimal choice in the next time slot. After taking action as per the policy, the agent is rewarded at time slot t + 1. To maximize its long-term discounted payoff, each agent aims to improve the correlation between states and actions. This motivates the agent to continually select the most favorable action during continuous interactions. The action and state at time t form the basis of the reward, which is defined as R for the pair $(s_t, a_t)$ according to Equation 4.3.

$$R = \sum_{t=0}^{T} \gamma^t r^t \tag{4.3}$$

Here, $\gamma^t$ denotes the discount rate, and $r^t$ means reward at time $t$. The efficiency of the RL agent relies on the design of the reward function.

**Policy:** A policy denoted as $\pi$ is a mapping from a state to an action. An agent uses policy to make decisions in an environment. It defines the behavior of the agent and guides its action depending on the present state of the environment. $\pi: q \rightarrow b$. The policy map provides probability *(Pr)* as specified in Equation 4.4.

$$\pi(q, b) = Pr\ (q_t = q, b_t = b) \tag{4.4}$$

### 4.3.1 Cutting-Edge Reinforcement Algorithms in Controlled Environment Agriculture

The application of RL in CEA has multiple advantages, such as enhanced productivity, decreased operational expenses, and sustainability. Reinforcement learning models thrive in responding to intricate and dynamic settings, including seasonal variations and equipment malfunctions. They can forecast agricultural health problems, enhance resource utilization, and guarantee continuous output throughout the year by perpetually learning and refining from their interactions. A high-quality model is not always necessary for RL to learn control strategies; it can learn them simply from data. Even with its ease of use and strong foundation, RL has limitations. The need for a specific and constrained set of states and control actions was the most important factor, especially for early RL algorithms. Under the guise of neural networks, function approximation was used to overcome the limitation. This section gives a comprehensive analysis of potential RL algorithms and their impact on implementing CEA.

Generally, RL algorithms are classified into value-based methods or policy-based methods [18]. Both offer distinct plans for making decisions in dynamic environments.

#### *4.3.1.1 Value-Based Reinforcement Learning Algorithm for Controlled Environment Agriculture*

Value-based RL algorithms learn the value function instead of deriving a policy directly. It assesses the anticipated cumulative reward of an agent's activities in specified states. The predominant algorithm in this area is Q-learning. The constraint in recalling state and action pairs, as it expanded, led to the expansion of DQN. Neural networks of DQN can approximate the value function instead of holding the Q-value for each state-action pair.

The idea of RL in any application is to give an immediate reaction to current environmental variable adjustments based on the agent's action prediction. Q-learning is a simple model-free algorithm in RL that works based on the Markov decision process. It defines behavior in each state so an agent can maximize its reward [19]. Q-learning is limited to a set of states and action spaces, so DQN becomes the most used RL algorithm. The DQN algorithm combines Q-learning and deep neural networks. It is designed to manage settings with complex or continuous state spaces.

It combines Q-learning with deep neural networks to handle environments with large or continuous state spaces.

This section analyses how DQN can be adopted in an irrigation system of CEA. To ensure enough hydration for the crops, it can be deployed to optimize water usage. The definition of DQN parameters in irrigation systems is given in Table 4.3.

Relying on the present Q-value and the $\epsilon$-greedy action-selection strategy, the DQN agent selects the action to be carried out at each time step *t*. $\epsilon$-greedy action selection is widely used in RL to strike a balance between exploitation and exploration. An algorithm for training an agent is given in *Algorithm 4.1*.

**Table 4.3** Deep Q-network parameters of the irrigation system in CEA.

| **Definition** | **Parameters** |
|---|---|
| States*(q)* | Humidity, Soil moisture level, Temperature, Water availability,<br>Crop growth height. |
| Action*(b)* | Decision on Increase/Decrease water flow |
| Reward*(r)* | Numerical reward signal based on plant growth, water adeptness, and penalties for overwatering. |
| Environment | The irrigation system operates in CEA. |
| Time*(t)* | - |
| $\epsilon$-greedy action selection | Random selection of action with probability $\epsilon$ or the highest Q-value action with probability $1$-$\epsilon$ |

**Algorithm 4.1** Agent training process of DQN.

Input: Humidity, Soil moisture level, Temperature, Water availability, Crop growth height.
Result: Optimized water usage policy
Initialize network Q with random weight $w_i$
Initialize replay memory D to the capacity C
for occurrence = 1 to R do
    receive initial state observation q
    for t = 1 to N do
        explore an action $b^t$ with probability 1-epsilon
        Or else exploit action $b^t$ =argmaxQ($q^t$,$b^t$,weight)
        run action $b^t$
        get reward $r^t$ for chosen action $b^t$ and next State $q^{t+1}$
        cache ($q^t$, $b^t$, $r^t$, $q^{t+1}$) into D
        Choose samples from D randomly
        Compute $y=r+\gamma b'\max Q(q',b';\theta)$
        Update the weights of the DNN for loss minimization through stochastic gradient descent
        Policy update $\pi(q^t)$ after every step
    end for
end for

### *4.3.1.2 Policy-Based Reinforcement Learning Algorithms in Controlled Environment Agriculture*

Unlike value-based RL algorithms, policy-based algorithms are more effective in handling continuous action spaces. This method directly learns the policy function without computing the value function. It directly maps state and action through policy optimization. Proximal policy optimization offers more stability and adaptability in CEA. It can optimize lighting, irrigation, and climate control systems in CEA.

The working procedure of policy-based RL comprises three parts. Defining state space, action space, and policy optimization. The parameters of the state space might be common in value and policy-based methods. The action space could be discrete or continuous. However, policy optimization makes the policy-based method stand out differently. The policy function guides the agent to elect an action that pays off a higher cumulative reward.

**Table 4.4** Types of policy-based RL algorithms.

| Algorithm | Suitability | CEA application |
|---|---|---|
| Policy gradient method | For direct optimization | Small-scale CEA system |
| Proximal policy optimization (PPO) | Ideal for handling multi-variable | Large vertical farming |
| Soft Actor-Critic (SAC) | Ideal for unpredictable dynamics | Fluctuating weather |
| Deterministic policy gradient | For continuous action spaces | Precise adjustments in irrigation or nutrient supply |

The policy is signified as $\pi(a|s;w)$ agent takes action '*a*' in state *s* with neural network parameter '*w*.' Over every state-action-reward cycle, the policy network is updated using parameter '*w*' to maximize the expected reward. Using gradient ascent, the policy is optimized. Various policy-based RL algorithms are listed in Table 4.4 along with their feasibility in CEA.

Recently, the complex CEA system demands combined advantages of value and policy algorithms. This hybrid method was coined as the Actor-Critic model. An actor-network selects an action based on policy. A critical network evaluates the policy based on the value function. It can concurrently acquire a Q-function (critic) and a policy (actor) [20].

### 4.3.2 Comparison of Policy and Value-Based Method

This comparison shows the advantages and disadvantages of each strategy, implying that the particular needs of the application, such as whether continuous or discrete actions are required. Based on six parameters, value-based and policy-based RL algorithms are compared using radar plots, as shown in Figure 4.4. The parameters are action space, efficiency, stability, scalability, easy implementation, and suitability for continuous actions. Value-based RL stays top in ease of implementation, efficiency, and stability. It is ideal for simple systems and discrete action scenarios. In contrast, policy-based RL is better suited for complex, dynamic situations, like those that demand precise control, because of its superior action space flexibility, scalability, and suitability for continuous actions.

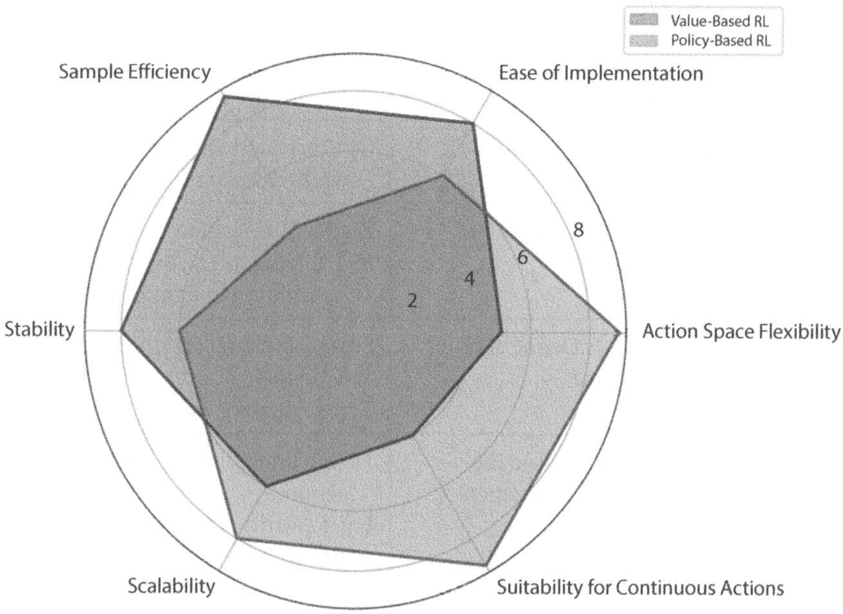


**Figure 4.4** Analysis of value-based and policy-based RL algorithms.

## 4.4 Case Study: Irrigation System Using Deep Q-Network

A case study on an irrigation system utilizing DQN would generally encompass the creation and implementation of an RL agent to enhance water distribution and utilization in the CEA system. Water scarcity constitutes a significant issue in certain regions of our nation. Effective water management is essential in the agricultural industry. Our irrigation system entirely depends on a predetermined schedule or human control. It results in the overutilization or underutilization of water. Integrating deep reinforcement learning, such as DQN, enables irrigation systems to dynamically adjust to crop needs, environmental factors, and soil moisture levels.

The methodology entails training a DQN agent to determine the time and quantity of irrigation depending on environmental variables [21]. Table 4.5 gives the working procedure breakdown, steps, and outcomes.

The DQN-based irrigation system realised water savings of 20–30% relative to traditional threshold-based techniques, while enhancing production by approximately 11% in certain instances.

**Table 4.5** The implementation method of DQN in the irrigation system.

| Working procedure | Steps | Description |
|---|---|---|
| Initial Setup | Step 1: Objective | Optimize agricultural output while reducing water usage. |
| | Step 2: Define state space | Feed real-time data.<br>E.g., temperature, soil moisture, humidity, and light intensity. |
| | Step 2: Define action space | Set up adjusting action through actuators.<br>E.g., regulating water |
| | Step 2: Define the reward function | Define immediate reward for an action.<br>E.g. $Reward = i * growthrate - j * Energycost - k * waterusage$<br>Where i, j, and k are tunable weights. |
| Algorithm Selection | Value-based algorithm | DQN |
| | Policy-based algorithm | Policy gradient method |
| | Hybrid algorithm | Deep deterministic policy gradient |
| Training Procedure | Step 1: Q-Value Initialization | Random values for all state-action pairs. (for DQN) |
| | Step 2: Action Selection | Exploration or exploitation<br>E.g., greedy action selection rule |
| | Step 3: Transition and feedback | • Implement the selected action<br>• Observe the new state and the reward. |
| | Step 4: Q-value update | $Loss=[r+\gamma a' maxQ(s',a';\theta)-Q(s\ a;\theta)]^2$<br>θ denotes DQN parameters. |
| Deployment | Real-time decision-making | The system adjusts dynamically to environmental changes. |

## 4.5 Challenges and Potential Solutions

Reinforcement learning contributes considerable value to CEA due to its ability to handle uncertainty effectively. More information regarding RL's contributions to CEA was included in this chapter. Still, some more challenges exist to overcome with deploying RL in practice. These problems encompass a wide range of factors, including environmental, operational, and technical [22, 23]. Let us start with the technical challenges in implementing RL in CEA.

### 4.5.1 Technical Challenges

This section discusses the technical challenges involved in RL.

- **Modelling a complex environment-** The modeling of a complicated environment that includes variations in temperature, humidity, and light is a very challenging endeavor. During the translation of trained models to real-world applications, imperfect simulations can result in policies that are less than ideal.
- **High-dimension state and action space-** The effectiveness of the RL model is dependent on the dimensionality of both the state and action space. Reinforcement learning algorithms are susceptible to being overwhelmed by high-dimensional state spaces. One example would be data collected in real-time by sensors that measure temperature, humidity, light, and $CO_2$ levels. Similarly, it is difficult to create an action space that strikes a compromise between the granularity of control and the computational feasibility of the space.
- **Delayed rewards-** In CEA, it is not possible to quickly analyze the impact of the actions taken by the agent. It is not possible to observe crop growth, which results in delayed reward signals in real-time utilization. This makes it more difficult for the agent to discover effective policies in a short time.
- **Training time and data requirements-** There is frequently a scarcity of high-quality datasets for the purpose of training RL models in CEA, particularly for crops or settings that are highly specialized. The learning process can be misled by noise in sensor data. For RL algorithms to learn successfully, they frequently require a significant length of training time and a substantial amount of data.

- **Generalization capabilities**- RL controllers must have a high degree of generalization to a variety of habitats and situations. Accomplishing this can be difficult, particularly in situations when the training environment is significantly different from the deployment environment.
- **Evaluation metrics**- To assess the performance of RL controllers in real-world settings, it is crucial to define relevant measures. However, given the dynamic and unexpected nature of these environments, this can be a tough task.

### 4.5.2 Environmental Challenges

- **Dynamic external influences**- Several external issues, such as power outages, fluctuating energy costs, and delays in the supply chain for resources, have the potential to disturb the assumptions of RL models. Real-time adaptation to these factors adds a layer of intricacy.
- **Heterogeneity in CEA systems**- The transfer of policies from one setting to another is made more difficult by the fact that alternative system designs, such as different greenhouse layouts or sensor combinations, are required.
- **Interdependencies among variables**- It is necessary to have advanced RL models to analyze the nonlinear and complicated relationships that exist between environmental parameters, such as the influence of humidity on temperature control.

### 4.5.3 Operational Challenges

- **Integration with existing systems**- A significant number of CEA configurations make use of legacy systems that were not intended for enhanced automation or AI integration, which makes the deployment of RL more difficult.
- **Experiment cost**- Training RL models in real-world CEA systems can be costly in terms of time, resources, and potential damage to crops.
- **Scalability**- To scale RL solutions across several CEA sites or a wide variety of crop kinds, it is necessary to have models that are both generalized and adaptable.

### 4.5.4 Potential Solutions

- **Hybrid Approaches-** It is possible to improve both efficiency and safety by combining RL with model-based methods or information about the domain. Methods such as transfer learning are used to modify previously learned models for use in different environments.
- **Robust Simulations-** The development of very accurate simulations reduces experimentation in the actual world. Incorporating stochastic aspects into simulations to capture the variability that exists in the real world.
- **Hierarchical and Modular RL-** By dividing jobs into smaller, more manageable subtasks handled by distinct RL agents, tasks can be broken down.
- **Safety Mechanisms-** To prevent harmful actions during training and deployment, the implementation of limits and safety checks is being considered.
- **Integration of IoT-** The utilization of IoT devices for real-time monitoring and feedback enables reinforcement-learning agents to make more informed decisions.

## 4.6 Reinforcement Learning Integration with Other Emerging Technologies

Reinforcement learning could greatly improve CEA when integrated with other cutting-edge technologies. Table 4.6 summarizes potential enhancements made through RL and emerging technology integration.

### 4.6.1 Adaptation Analysis of Cutting-Edge Technologies in Controlled Environment Agriculture

Figure 4.5 is a comparative plot depicting the adoption rates of several technologies in CEA. The data illustrates a prevailing trend, emphasizing the significance of IoT, ML, and other promising technologies such as digital twins and RL.

#### *4.6.1.1 Machine Learning*

The application of ML in agriculture has been proven successful in optimizing several processes in the agricultural sector. These processes include

**Table 4.6** Integrating RL with emerging technologies.

| Technology | Use case | Reinforcement learning application |
|---|---|---|
| Internet of Things | Precision farming | Use IoT data to optimize the process. |
| Artificial intelligence and computer vision | Disease detection | Learn to identify diseases and recommend interventions. |
| Robotics | Smart farming robots | Train the robots to navigate on farm layout. |
| Drones | Aerial monitoring and spraying | Decide where and how to spray pesticides |
| Blockchain | Supply chain optimization | Optimize logistics and storage |
| Digital twins | Virtual farm | Simulate strategies like irrigation patterns or crop rotations within the digital twin |
| Big data analytics | Weather-based decision-making | Adapt planting schedules or resource allocation |

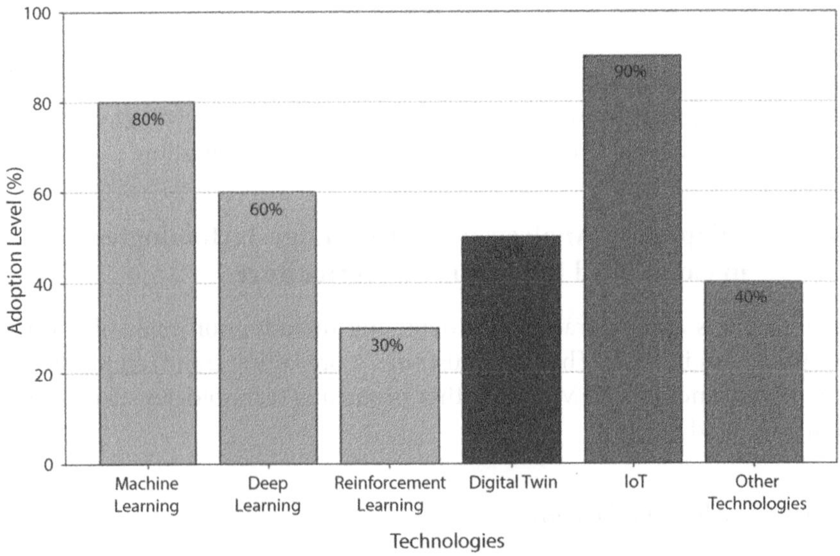


**Figure 4.5** Adoption of technologies in CEA.

predictive yield modeling, disease detection, precision spraying, and market forecasting. To improve decision-making and minimize resource waste, these applications mainly rely on data collected *via* IoT sensors. These sensors offer insights into crop health and environmental conditions.

#### *4.6.1.2 Deep Learning*

Deep learning models, particularly convolutional neural networks, are frequently employed for image-based tasks such as disease diagnosis, pest detection, and crop quality evaluation. Utilizing this method, one can get a higher level of precision when identifying patterns within agricultural records.

#### *4.6.1.3 Internet of Things and Sensors*

The IoT has revolutionized agriculture by monitoring environmental elements in real-time. Incorporating ML models into management techniques makes it easier to implement adaptive management strategies for irrigation and resource optimization.

#### *4.6.1.4 Digital Twin*

The term "Digital Twin" refers to the creation of a virtual model of the farm environment. This model enables the modeling and analysis of a variety of scenarios, resulting ultimately in optimal resource allocation and predictive management. The CEA operations are improved by the utilization of this technology, which incorporates sensor data and ML algorithms [24, 25].

#### *4.6.1.5 Reinforcement Learning*

Fewer studies have investigated the implementation of RL algorithms in CEA, with a particular emphasis on the applicability and efficiency of the algorithms. Figure 4.6 provides a high-level summary of the RL algorithms that have been utilized in a variety of research. Although RL models are not as prevalent in the existing literature, they have the potential to be utilized for the dynamic optimization of irrigation systems and autonomous greenhouse management [26–28].

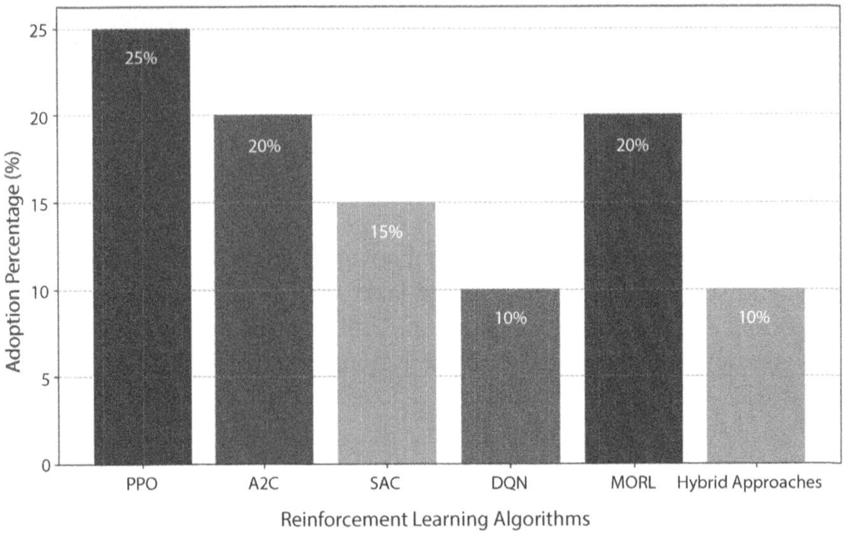


**Figure 4.6** Adoption of RL algorithms in CEA.

## 4.7 Conclusion

By offering intelligent, adaptive, and resource-efficient systems capable of addressing crucial difficulties in modern farming, RL was poised to revolutionize CEA. The purpose of this chapter is to demonstrate the revolutionary potential of cutting-edge RL algorithms in terms of optimizing resource utilization, increasing agricultural output, and responding to changing environmental conditions. This study bridges the AI and agricultural technology domains in providing a comprehensive framework for leveraging RL to overcome some traditional farming limitations. As such, it will help researchers and practitioners better understand this field, including its challenges and opportunities. This will inspire further innovation and research to pave the way toward sustainable and intelligent agricultural practices that support economic and environmental goals.

**AI disclosure statement:** The author confirms that no artificial intelligence (AI)-generated content or AI-assisted writing tools were used in the preparation of this chapter (including summaries and the abstract). All text, analysis, and conclusions are the result of the author's original work.

## References


1. Rane, J., Kaya, O., Mallick, S.K., Rane, N.L., Smart Farming Using Artificial Intelligence, Machine Learning, Deep Learning, and ChatGPT: Applications, opportunities, challenges, and future directions. *Gen. AI. Agric. Educ. Bus.*, 41, 218, 2024.
2. Sasmal, B., Das, G., Mallick, P., Dey, S., Ghorai, S., Jana, S., Jana, C., Advancements and challenges in agriculture: A Comprehensive Review of Machine Learning and IoT Applications in Vertical Farming and Controlled Environment Agriculture. *Big Data Com. Vis.*, 4, 67, 2024.
3. Nesteruk, S., Shadrin, D., Pukalchik, M., Somov, A., Zeidler, C., Zabel, P., Schubert, D., Image compression and plants classification using Machine Learning in controlled-environment agriculture: Antarctic station use case. *IEEE Sens. J.*, 21, 17564, 2021.
4. Ragaveena, S., Shirly Edward, A., Surendran, U., Smart Controlled Environment Agriculture Methods: A holistic review. *Rev. Env. Sci. Bio/Tech.*, 20, 887, 2021.
5. Leskovar, D.I., Choi, S., Harvey, J.T., Lee, C., Nagila, A., Niu, G., Dash, P.K., Controlled Environment Agriculture Trends in Texas: Challenges, opportunities, and research approaches, in: *IX South-Eastern Eur. Sym. on Veg. and Pot.*, vol. 1391, p. 197, 2023.
6. Ojo, M.O. and Zahid, A., Deep Learning in Controlled Environment Agriculture: A Review of Recent Advancements, Challenges and Prospects. *Sensors.*, 22, 7965, 2024.
7. Mohamed, M., Agricultural Sustainability in the Age of Deep Learning: Current Trends, Challenges, and Future Trajectories. *Sustain. Mach. Intell. J.*, 4, 2, 2022.
8. Bu, F. and Wang, X., A smart agriculture IoT system based on deep Reinforcement Learning. *Future Gen. Comput. Sys.*, 99, 500, 2019.
9. Gautron, R., Maillard, O.A., Preux, P., Corbeels, M., Sabbadin, R., Reinforcement learning for crop management support: Review, prospects and challenges. *Comput. Electron. Agric.*, 200, 107182, 2022.
10. Zhang, Z., Boubin, J., Stewart, C., Khanal, S., Whole-Field Reinforcement Learning: A fully autonomous aerial scouting method for precision agriculture. *Sensors.*, 20, 6585, 2020.
11. Din, A., Ismail, M.Y., Shah, B., Babar, M., Ali, F., Baig, S.U., A deep Reinforcement Learning-based multi-agent area coverage control for smart agriculture. *Comput. Electr. Eng.*, 101, 108089, 2022.
12. Madondo, M., Azmat, M., Dipietro, K., Horesh, R., Jacobs, M., Bawa, A., O'Donncha., F.A., Swat-Based Reinforcement Learning Framework for Crop Management, *arXiv preprint arXiv:2302.04988*, 2023.
13. Turchetta, M., Corinzia, L., Sussex, S., Burton, A., Herrera, J., Athanasiadis, I., Krause, A., Learning long-term crop management strategies with cycles-gym. *Adv. Neu. Info. Pro. Sys.*, vol. 35, p. 11396, 2022.

14. Alibabaei, K., Gaspar, P.D., Assunção, E., Alirezazadeh, S., Lima, T.M., Irrigation optimization with a deep Reinforcement Learning model: Case study on a site in Portugal. *Agric. Water Manag., 263*, 107480, 2022.
15. Binas, J., Luginbuehl, L., Bengio, Y., Reinforcement learning for sustainable Agriculture, in: *ICML 2019 Wor. Cli. Change: How Can AI Help*, vol. 43, p. 1254, 2019.
16. Gorantla, S. and Veluchamy, S., Reinforcement Empowered Deep Network Enabled Optimal Controlled Environment for Crop Yield, in: *2024 15th Int. Con. on Comp. Com and Net. Tech.*, vol. 23, p. 6, 2024.
17. Hao, Z., Li, X., Meng, C., Yang, W., Li, M., Adaptive spraying decision system for plant protection unmanned aerial vehicle based on Reinforcement Learning. *Int. J. Agric. Bio. Eng.*, 15, 16, 2022.
18. Shakya, A.K., Pillai, G., Chakrabarty, S., Reinforcement Learning algorithms: A brief survey. *Exp. Sys. App.*, 231, 120495, 2023.
19. Udayakumar, K. and Ramamoorthy, S., Intelligent Resource Allocation in Industrial IoT using Reinforcement Learning with Hybrid Meta-Heuristic Algorithm. *Cyber Sys.*, 54, 1241, 2022.
20. Dong, R., Du, J., Liu, Y., Heidari, A.A., Chen, H., An enhanced deep deterministic policy gradient algorithm for intelligent control of robotic arms. *Front. Neuro. Info.*, 17, 1096053, 2023.
21. Alibabaei, K., Gaspar, P.D., Assunção, E., Alirezazadeh, S., Lima, T.M., Soares, V.N.G.J., Caldeira, J.M.L.P., Comparison of On-Policy Deep Reinforcement Learning A2C with Off-Policy DQN in Irrigation Optimization: A Case Study at a Site in Portugal. *Comp.*, 11, 104, 2022.
22. Wang, Z. and Hong, T., Reinforcement learning for building controls: The opportunities and challenges. *App. Energy*, 269, 115036, 2020.
23. Dulac-Arnold, G., Levine, N., Mankowitz, D.J., *et al.*, Challenges of real-world Reinforcement Learning: definitions, benchmarks, and analysis. *Mach. Learn.*, 110, 2419, 2021.
24. Chaux, J.D., Sanchez-Londono, D., Barbieri, G., A Digital Twin Architecture to Optimize Productivity within Controlled Environment Agriculture. *Appl. Sci.*, 11, 8875, 2021.
25. Goldenits, G., Mallinger, K., Raubitzek, S., Neubauer, T., Current applications and potential future directions of Reinforcement Learning-based Digital Twins in agriculture, *arXiv preprint arXiv:2406.08854*, 2024.
26. Liu, H., Zhang, J., Zhou, Z., Dai, Y., Qin, L., A Deep Reinforcement Learning-Based Algorithm for Multi-Objective Agricultural Site Selection and Logistics Optimization Problem. *Appl. Sci.*, 14, 8479, 2024.
27. Gandhi, R., Deep Reinforcement Learning for Agriculture: Principles and Use Cases. *Data Sci. Agri. Nat. Res. Manag.*, 43, 75, 2022.
28. Lu, Y., Gong, M., Li, J., Ma, J., Optimizing controlled environmental agriculture for strawberry cultivation using RL-informer model. *Agro.*, 13, 2057, 2023.

5

# Augmented Reality-/Virtual Reality-Assisted Deep Reinforcement Learning-Based Model toward Management of Soil Microbes on Organic Farms

**G. Amuthavalli[1]*, U. Palani[2], G. Vallathan[3] and Prasanth Aruchamy[4]**

[1]*Department of Electronics and Communication Engineering, Sri Venkateshwaraa College of Engineering & Technology, Affiliated to Pondicherry University, Puducherry, India*

[2]*Department of Electronics and Communication Engineering, IFET College of Engineering, Villupuram, Tamil Nadu, India*

[3]*Department of Electrical and Electronics Engineering, Geethanjali College of Engineering and Technology, Hyderabad, India*

[4]*Department of Computer Science and Engineering, Vel Tech Rangarajan Dr. Sagunthala R&D Institute of Science and Technology, Chennai, Tamil Nadu, India*

## *Abstract*

The potential implementation of sustainable agriculture under climatic shocks necessitates adopting a holistic organic farming approach, in which the practice of organic soil amendments with technological advancements further reassures the sustainability and food security despite environmental challenges. In the context of climate-smart agriculture, optimal organic farming strategies are achieved by promoting a thriving ecosystem of soil microbes, for which the artificial intelligence manages microbial activity and dependent factors. Ecotoxicological risks due to urbanization are a major threat to soil microbes and affect soil health, impacting crop yield. In this chapter, a deep reinforcement learning-based model with the assistance of augmented reality/virtual reality is discussed to manage soil microbes in organic farming, which effectively assesses ecotoxicological risks

**Corresponding author*: amuthavalli.phd1@gmail.com

Anitha Velu, Prasanth Aruchamy, Raghu Ramamoorthy, Rajesh Kumar Dhanaraj and Seifedine Kadry (eds.) Climate Smart Agriculture, (85–100) © 2026 Scrivener Publishing LLC

in the soil and provides the best practice of climate-smart agriculture to achieve sustainability.

***Keywords*:** Soil microbes, deep reinforcement learning, augmented reality/virtual reality, climate, organic farming

## 5.1 Introduction

To build a socio-economically balanced food system, sustainable agriculture becomes a necessity due to climate change and resource scarcity. The United Nations' Sustainable Development Goals (SDGs), like Zero Hunger, Clean Water and Sanitation, Decent Work and Economic Growth, Responsible Consumption and Production, Climate Action, and Life on Land, place strong emphasis on sustainable agriculture amid critical global threats like water scarcity, climate change, biodiversity loss, and soil degradation [1]. The unpredictable impacts of climate change have triggered ecological degradation, as traditional farming methods often rely heavily on synthetic pesticides and chemicals, as well as single-crop systems [2]. To combat pressing global challenges, sustainable agriculture promotes soil enrichment through organic farming [3], where soil microbes enhance robust crop health and productivity. However, the natural processes and biological interactions of soil microbes are affected by climate change, such as erratic weather, prolonged droughts, floods, elevated temperatures, and declining soil fertility. The critical role of soil microbial groups is to maintain soil fertility by focusing on nutrient cycling, decomposing organic matter, suppressing diseases, and rigid soil structures. Climatic fluctuations such as increased temperatures, erratic precipitation, and the effects of global warming can accelerate the metabolism of microbes, reducing microbial diversity, disrupting the water-holding capacity of the soil and interactions of microbes, altering the natural freeze-thaw cycles, and imposing stress on the soil biota [4]. In organic farming, soil amendments play a significant role in enhancing plant health and ecologically balanced productivity, thereby promoting climate-resilient and sustainable agriculture. Common organic soil amendments include compost, green manure, vermicompost, animal manure, crop residues, and mulches [5]. Figure 5.1 depicts organic soil amendments to enrich soil fertility in organic farming to promote sustainable agriculture amid climate change. Compost improves aeration and water retention and releases nutrients slowly, supported by decomposed plant or animal waste [6]. Vermicompost, rich in enzymes and nutrients, enhances soil fertility through enzymatic activity [7]. Green manures

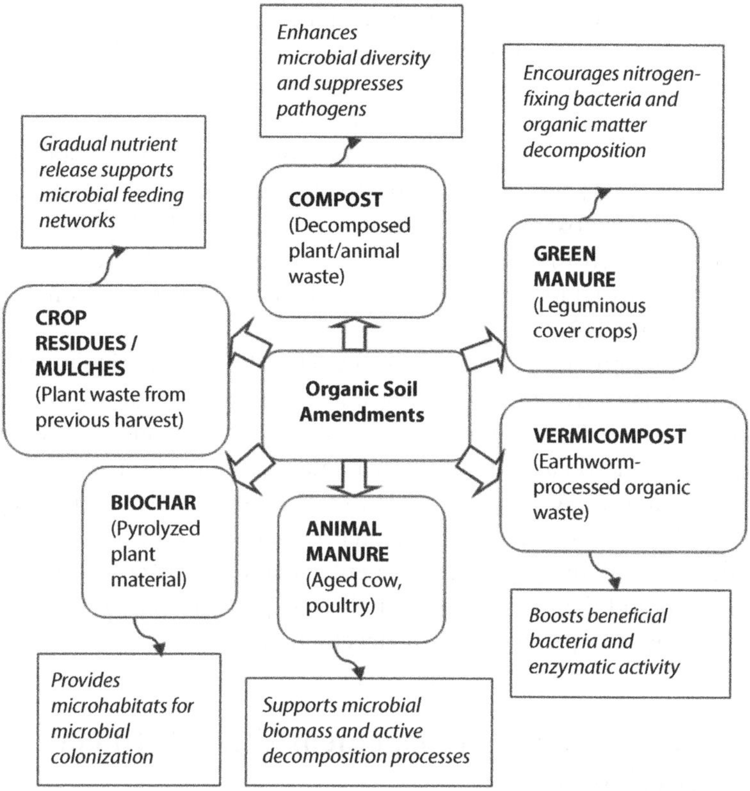


**Figure 5.1** Organic soil amendments in organic farming.

improve the texture and reduce soil erosion, thereby increasing nitrogen content [8]. Animal manure supports biomass production by supplying macro- and micronutrients. Biochar improves the water-holding capacity and soil porosity, creating microhabitats for microbes [9]. Crop residues or mulches help regulate temperature and retain soil moisture, thereby preventing soil erosion. However, climate change adversely affects the effectiveness of organic soil amendments, necessitating adaptable and optimized soil amendment strategies [10].

The microbial populations are adversely affected by climate change, therefore making it difficult to monitor and regulate real-time soil microbial activities without advanced tools. The imposing challenges include inadequate awareness of microbial management practices, lack of decision-making supporting frameworks, limited resources for monitoring, ecotoxicological risks, and unpredictable outcomes [11]. The most significant and least addressed challenge is the ecotoxicological risk, which includes contaminants like heavy metals, microplastics, synthetic pesticides and herbicides, antibiotic residues, and organic pollutants leading to altered functional profiles of microbial activity, disrupted nutrient cycling, and reduced biodiversity [12]. A soil health monitoring system, along with mitigation of ecotoxicological risks, is the solution to manage soil microbial activity. Thus, the overall impacts of climate change, unsustainable practices, and ecotoxicological risks demand a multi-disciplinary system-oriented approach integrating artificial intelligence (AI) and augmented reality (AR)/virtual reality (VR), which offer a transformative solution for organic farming [13].

The chapter provides deep insights into climate-smart agriculture (CSA) from the perspective of integrating emerging technologies into organic farming. Chapter 5.1 introduces sustainable agriculture and challenges in maintaining microbial activity, and Chapter 5.2 focuses on soil microbial management using AI. Chapter 5.3 emphasizes the significance of integrating AR/VR in organic farming. Chapter 5.4 elaborates an AR/VR-assisted AI model for soil microbial management. Chapter 5.5 discusses real-world applications and their challenges. Finally, Chapter 5.6 concludes the chapter with a ray of hope for futuristic development in CSA.

## 5.2 Soil Microbial Management Using Artificial Intelligence

Artificial intelligence is transforming soil microbial management by enabling precise analysis, prediction, and optimization of soil health. Advanced AI techniques, such as machine learning (ML) and deep learning (DL), analyze microbial interactions, predict nutrient cycles, and optimize soil conditions for sustainable agriculture [14]. The AI-driven models, including support vector machines (SVM) and artificial neural networks, assess microbial diversity, organic matter decomposition, and soil fertility with high accuracy [15, 16]. These innovations facilitate real-time soil monitoring, data-driven decision-making, and precision farming. By integrating AI with Internet of Things (IoT) sensors and big data analytics, researchers can enhance soil microbial activity, improving crop productivity and environmental sustainability [17]. The datasets of microbial communities give a deeper understanding of the biological processes underlying soil health and fertility. The ML technique utilizes massive data of soil microbes generated by high-throughput sequencing approaches/methods such as gene sequencing, metagenomics, and metatranscriptomics.

The high-dimensional data of microbes contain enormous operational taxonomic units, each of which has relative data, genetic markers, and metadata of environmental changes like soil pH, moisture, temperature, and organic content [18]. The conventional statistical methods struggled with their limited ability to model the non-linear intricacies between them. Fortunately, AI models can enable soil categorization by microbial health and predict dominant taxa in response to environmental factors associated with plant health or disease. More specifically, supervised ML algorithms, such as SVMs, gradient boosting machines, and random forests, are effective in identifying the complex patterns from high-dimensional data. The feature selection in AI influences soil fertility and crop performance by automatically identifying microbial features, facilitating the management of microbes by applying biofertilizers or compost amendments. The feature selection algorithms, such as recursive feature elimination, least absolute shrinkage and selection operator regression, and Shapley additive explanations, identify the microbial biomarkers that effectively reflect soil health status [19]. Unsupervised learning techniques are employed to facilitate the community dynamics, detection of anomalies such as sudden declination in microbial diversity, and clustering samples with similar microbial ecologies [20]. These are crucial in revealing spatial and temporal patterns in organic farms. The clustering algorithm along with dimensionality reduction techniques aid in profiling the microbial community structures in the absence of predefined labels. The AI models like DL models learn the complexity in relating the microbial taxa and environmental parameters, and are explored and trained to suggest specific microbial enrichment strategies. The multimodal data fusion, in which the AI models facilitate the integration of the microbial data with other sensor data to analyze the microbiome, allows prediction of the microbial response to environmental variables. The real-time AI-powered dashboards can provide early warnings for imbalanced microbial activity, recommend remediation measures and visualize the impacts through user-friendly interfaces.

## 5.3 Integration of Augmented Reality and Virtual Reality in Organic Farming

A transformative shift in organic farming happens by integrating immersive technologies such as AR and VR and redefining the perspective of handling biological and ecological data [21]. These technologies enhance the farmer's ability to make real-time, evidence-based decisions, while seamlessly bridging the gap between physical and digital realms and empowering the precision management of soil microbial health, which is the ensuring factor of sustainable organic agriculture. The need for AR and VR in agriculture is to visualize real-time data such as weather conditions and patterns, soil health, crop growing nature, pest control, irrigation, and fertilization; thereby, precise decision-making becomes possible, reducing human intervention and risks. Also, simulation offers training for farmers on all the crop field-oriented activities and improving the efficiency with optimal use of resources [22]. Farmers can plan different farming scenarios and their yield, with great transparency, which can improve market opportunities. As depicted in Figure 5.2, the AR and VR can be integrated to empower more sustainable, efficient, and resilient organic farming practices. The AR mainly supports real-time monitoring of microbiomes and effective decision-making, which it leverages into organic farming in the following activities: 1) Real-time visualization of soil microbiome, 2) Precision intervention, and 3) Crop compatibility based on the microbes. The sensor system in the crop field is connected with the AR-based system. The data from the sensors, such as soil pH, temperature, moisture, microbial $CO_2$ respiration, and redox potential, are collected, and an interpretation is done by AI models. Then, the collected

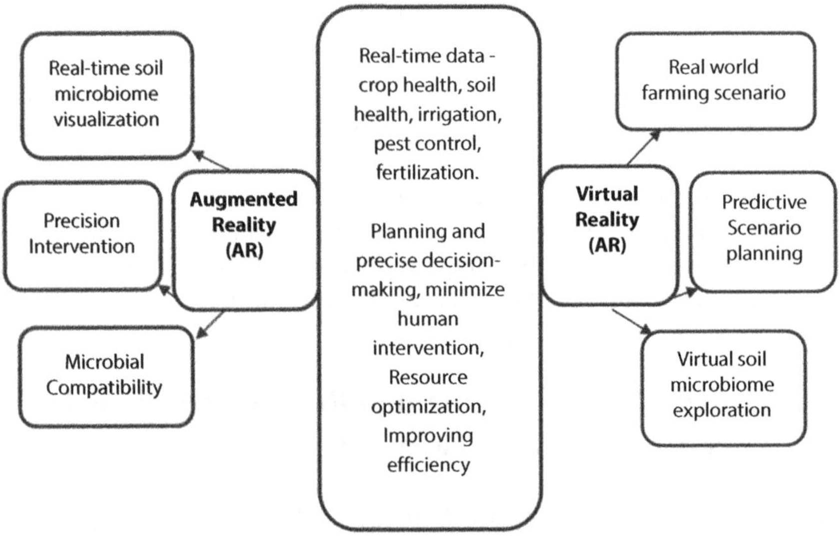


**Figure 5.2** Integration of AR and VR in organic farming.

data are spatially projected onto the field maps by the AR devices, facilitating farmers in observing zones of microbial activity, potential pathogens, and diversity indices as color-based coded heatmaps [23].

The global positioning system (GPS), along with the AR system, guided farmers in applying the exact needed tailored organic amendments in real-time. Also, integrating deep reinforcement learning (DRL) models enhances the microbial management strategies based on the climatic conditions and environmental feedback to provide a good yield outcome [24]. The crop compatibility can be observed by suggesting crop rotations or a plan of intercropping by AR systems, which mainly promotes beneficial microbes like mycorrhizae with vegetables and rhizobia with legumes. The system also supports farmers in visualizing the compatibility of microbes for different crop species, which are layered over sections of soil. Virtual reality is integrated to utilize immersive environmental experiences through training modules, predictive analysis through predictive scenario planning, and virtual exploration of the soil microbiome. Transitioning or new organic farmers can understand the microbial interactions, soil biology concepts, biofertilizer applications, and composting techniques by exploring the interactive VR environments. The training modules are used to learn the cause-and-effect scenarios, such as over-irrigation on denitrifying bacteria or tillage impacts on microbial diversity. The farmer can virtually experiment by applying fermented plant extracts in one season will affect the richness of microbes and resilience of the crop in the other season. The DRL-based models are used to simulate the impact of different organic management practices with respect to time. The expected outcomes of organic farming such as nitrogen mineralization, disease suppression, and carbon sequestration are simulated virtually, thereby farmers are trained on the organic amendments and practices. The VR headset gives an immersive experience of observing microscale visualization of the microbiome and also exploring the microbial world through animated representations of enzymatic activities, microbial food webs, and behavior of the microbial colony based on metagenomic data and lab sequencing [25]. The potential of AR/VR integration provides the benefits of time and resource efficiency, capacity building, enhanced data interpretation, precision and personalization, and sustainability and resilience. The integration of AR and VR succeeded only in coupling them with DRL models and providing a continuously evolving decision support system.

## 5.4 Augmented Reality-/Virtual Reality-Assisted Deep Reinforcement Learning Model for Soil Microbial Management

### 5.4.1 Framework of Augmented Reality-/Virtual Reality-Assisted Deep Reinforcement Learning-Based Model

In organic farming, long-term agricultural productivity is achieved by sustaining soil fertility and plant health without relying on synthetic inputs, but by effectively managing soil microbial activity. However, the dynamic microbial ecosystem and spatial variability pose challenges for effective management and demand a data-based, interactive, and intelligent approach. The novel, immersive framework that enhances the decision-making process is AR-/VR-assisted DRL model that has a layered

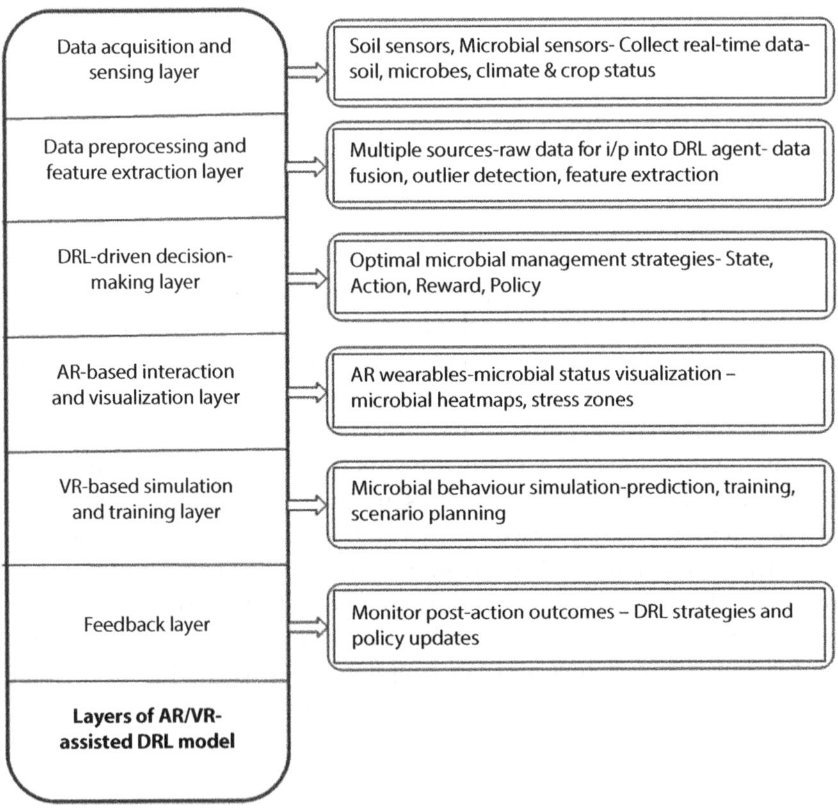


**Figure 5.3** Framework of AR/VR-assisted DRL-based AI model.

framework architecture with the following six layers as depicted in Figure 5.3: 1) Data acquisition and sensing layer; 2) Data preprocessing and feature extraction layer; 3) DRL-driven decision-making layer; 4) AR-based interaction and visualization layer; 5) VR-based simulation and training layer; and 6) Feedback layer. In the data acquisition and sensing layer, sensors play a major role in providing soil-based data, such as pH, temperature, moisture, $CO_2$ flux, microbial data, such as rainfall, solar radiation, weather forecasts, and crop parameters, such as root biomass, crop growth stages, and canopy cover [26]. The real-time data are collected and sent to the preprocessing layer, where data from multiple sources are fused by detecting outliers and aligning with the time frame.

Features, such as microbial richness, enzymatic activity, are extracted, and the raw data are given as input to DRL agents. In the DRL-based decision-making layer, four components, such as state space, action space, reward function, and policy network, are used, where multivariate environmental and microbial observations, organic interventions, encouragement of microbial balance, pathogen suppression, nutrient cycling, and optimal decision-making are done [27]. The AR wearables in the AR-based interaction and visualization layer are used to visualize the microbial heatmaps, stress zones, geospatially anchored guidance, and interactive control [28]. In the VR-based simulation and training layer, microbial behaviour is simulated to predict the soil microbial dynamics, educating farmers on composting impacts, microbiome behavior, and beneficial microbes. The feedback layer monitors post-action outcomes such as crop condition, soil texture, $CO_2$ respiration improvement, and updates the DRL policy based on reward signals and dynamic conditions.

### 5.4.2 Soil Contamination Identification by Augmented Reality Visualization

Soil microbial equilibrium is the backbone of sustainable organic farming and supports productivity, aligning with climate-smart agriculture goals. Disruption of soil microbial equilibrium by contaminants such as excessive salts, pathogenic microorganisms, heavy metals, and pesticide residues reduces agricultural productivity [29]. Traditional methods like laboratory-level chemical assays or satellite-supported imaging techniques are non-spatially resolved at the micro-farm level, have delayed responses, and are difficult for farmers to interpret [30]. The existing gap can be addressed to enable real-time, spatially explicit identification and interactive remediation using AR-based visualization systems. The AR wearables enhance farmers' perceptions by providing virtual experiences of a physical farm environment [31]. An intuitive, immersive decision-support experience can be obtained by integrating AR interfaces, field-sensing technologies, and ML-based contamination detection. Contaminant heavy metals like Pb, Cd, As, and Cr from industrial waste and sewage sludge damage DNA, inhibit microbial enzymes, and reduce biodiversity. Pesticide residues from conventional use disrupt microbial communities, whereas the alkalinity and salinity of the water due to over-irrigation, poor drainage, and saline water inhibit the N-fixing bacteria and reduce microbial respiration. The over-application of manure or compost results in leaching of nutrients, anaerobic zones, and proliferation of pathogens. The framework of an AR-based contaminant identification system has the following layers: 1) Data acquisition layer; 2) contamination detection layer; 3) AR interface layer; 4) interaction and visualization layer; and 5) feedback layer. The sensors provide soil-based data such as pH, EC, temperature, moisture, redox potential, and a few other advanced sensing elements like e-nose sensors, handheld X-ray fluorescence analyzers, calorimetric test kits, drone or satellite imagery systems, which are used to collect the data of soil, crop, and climate. Both ML and DL models with classification techniques such as a convolutional neural network, random forest, or anomaly detection classify contaminated ones and are trained on sensor fusion data, spectral data, and historical contamination patterns [32]. The AR interface layer has AR wearables to support the interaction and visualization layer, where the contaminant type, level, and remedial suggestion will be handled based on real-time sensor input. The feedback layer tracks changes in the status of contamination over time and informs the system to update the ML models. The combined AR-DRL feedback ensures the spatially and ecologically optimized microbial restoration. The multisource integration, field-level spatial granularity, farmer inclusivity, real-time feedback, and cost-effective long-term usage of AR-based soil contamination visualization provide data fusion from different sources, micro-zone remediation, planting decisions, visual literacy, immediate decision-making, and avoidance of frequent lab-based contamination analysis.

### 5.4.3 Virtual Reality Simulation-Based Prediction of Microbial Response to Contaminants

The soil health of organic farming highly relies on soil microbial communities, which drive nutrient cycling, decompose organic matter, resist diseases, and sequester carbon content. The function of microbial communities is disrupted by contaminants such as heavy metals, pesticide residues, salinity, and alkalinity, becoming a threat to organic farming. The functional resilience of microbial communities is challenging, but it can be resolved by predicting the microbial responses to contaminants using simulations of VR, where ecological models and AI offer an immersive platform for understanding the dynamics of microbial activity. By visualizing and interacting with soil microbiomes in a simulated environment, VR enhances farmers' decision-making ability on microbial interventions in contaminated zones. On integrating DRL models with VR simulations, an understanding of real-time data and predicted outcomes is evolved, and acts as an active decision-support system. The framework for VR simulation of microbial contaminant response comprises the following layers: 1) Data integration layer; 2) predictive modeling layer; and 3) VR visualization and immersive interface layer. The data integration layer aggregates and preprocesses data from different sources such as soil sensors, contaminant measurements, microbial sequencing, and historical data patterns. The collected and processed data are used to predict the algorithms and support 3D visualizations. In the predictive modelling layer, AI/ML models like DRL models are used with techniques such as neural networks, random forest, and SVM to train on the shifting nature of microbial community due to contaminants, observing the enzyme activity reduction patterns, and to be resilient with functional redundancy. On learning the optimal remediation strategies from virtual interventions, microbial stability is maximized over time. In the VR visualization and immersive interface layer, 3D soil and root zones with animated microbial entities are used to understand features such as microbial population dynamics in contaminated and remediated states, colour code-based stress markers, and time-lapse microbial growth under different interventions. The immersive interface simulates different scenarios by tracking microbial richness or diversity metrics under various levels of contamination and virtual DRL-based predicted outcomes across simulated climatic conditions.

## 5.5 Real-World Applications and Their Challenges in Augmented Reality-/Virtual Reality-Assisted Organic Farming

Both AR and VR provide a paradigm shift in organic farming by effectively improving farmers' decision-making, optimal resource management, and sustainability. Augmented reality offers real-time overlays for soil health monitoring, precision irrigation, and pest control. Virtual reality, on the other hand, enables immersive training and virtual farm simulations for farmers. These technologies improve efficiency, reduce environmental impact, and ensure better crop yields. However, real-world applications face challenges such as high implementation costs, limited access to advanced technology in rural areas, data integration issues, and the need for farmer training. Overcoming these barriers is crucial for widespread adoption and long-term success in smart organic farming. Traditional agricultural methods have long relied on manual observation, experience-based decision-making, and generalized farming practices. Farmers assess soil health through periodic lab testing and visual inspections, which can be time-consuming and sometimes inaccurate. Pest and disease control often involves the uniform application of chemical pesticides, leading to potential overuse and environmental concerns. Similarly, irrigation follows fixed schedules rather than adjusting to real-time soil moisture and weather conditions, resulting in water inefficiencies. Crop yield predictions in conventional farming are based on historical trends rather than real-time data, making them less precise. These methods also require significant labor and higher resource utilization, leading to increased costs and inefficiencies. Additionally, conventional farming lacks adaptability to climate change, making it vulnerable to extreme weather conditions. Sustainability is another challenge, as excessive use of fertilizers and pesticides can degrade soil quality and harm ecosystems. Furthermore, farmer training is primarily hands-on and requires years of experience to develop expertise. Artificial intelligence-driven approaches revolutionize agriculture by incorporating data analytics, automation, and real-time monitoring to optimize farming practices [33]. Decision-making is enhanced through AI-powered predictive models that analyze environmental data to provide precise recommendations for soil management, pest control, and irrigation. Artificial intelligence-based soil health management uses IoT sensors and ML algorithms to continuously monitor soil conditions, ensuring balanced nutrient levels. Smart pest detection systems utilize computer vision and AI algorithms to identify early signs of infestations, enabling targeted and eco-friendly pest control measures. Artificial intelligence-driven precision irrigation systems analyze weather patterns and soil moisture levels to optimize water usage, reducing waste. Advanced ML models accurately predict crop yields based on real-time environmental conditions, helping farmers plan better. These approaches also minimize costs by reducing labor, water, and fertilizer usage. Artificial intelligence-integrated agriculture is highly adaptable to climate change, using ML to dynamically adjust to environmental variations. Additionally, AI-powered AR/VR training modules provide immersive learning experiences, allowing farmers to adopt modern farming techniques more efficiently. In contrast, AI-integrated approaches enhance efficiency, sustainability, and adaptability by leveraging automation, real-time data, and predictive analytics. By transitioning to AI-driven smart farming, agriculture can become more resilient, productive, and environmentally sustainable. However, the limitations are high implementation costs, limited access to technology, digital literacy and farmer training, data collection and accuracy issues, scalability challenges, maintenance and technical support, privacy and data security concerns, limited localized AI models, dependency on external vendors, and farmers' hesitation to change.

## 5.6 Conclusion and Future Prospects

Climate-smart agriculture is an integrated approach that enhances agricultural productivity while ensuring sustainability and resilience to climate change. Key findings highlight that CSA improves soil health, water conservation, and biodiversity by incorporating sustainable practices such as organic farming, precision agriculture, and agroforestry. Artificial intelligence-driven technologies, including DRL and AR/VR, further optimize resource management and decision-making. CSA significantly reduces greenhouse gas emissions through efficient land-use practices, minimal chemical input, and carbon sequestration techniques. The adoption of AI-based risk assessment models helps predict and mitigate the impact of extreme weather events, ensuring food security and climate resilience. Additionally, CSA enhances livelihoods by promoting adaptive farming techniques, enabling small-scale farmers to reduce production risks while improving crop yields. However, challenges such as high costs, limited access to technology, and the need for farmer education hinder large-scale adoption. Addressing these barriers through affordable AI solutions, government support, and knowledge-sharing initiatives is crucial. Overall, CSA presents a sustainable pathway toward climate-resilient agriculture, ensuring long-term environmental conservation and food security. Future research in CSA should enhance AI-driven decision-making, precision farming, and sustainability to improve resilience against climate change. Advancements in DRL, AR/VR, and IoT can optimize soil microbial management, resource allocation, and risk assessment. Research should also focus on affordable, accessible technologies for small-scale farmers while reducing AI model biases through diverse, region-specific datasets. Long-term impact assessments are needed to evaluate CSA's effectiveness in climate mitigation, biodiversity enhancement, and food security. Interdisciplinary studies integrating climate science, agronomy, and AI can drive innovations in carbon sequestration, sustainable irrigation, and eco-friendly pest control. Addressing policy frameworks, ethical concerns, and farmer education will support widespread adoption. Future research should prioritize scalable, cost-effective, and environmentally sustainable CSA models that benefit both farmers and ecosystems.

**AI disclosure statement:** The author confirms that no artificial intelligence (AI)-generated content or AI-assisted writing tools were used in the preparation of this chapter (including summaries and the abstract). All text, analysis, and conclusions are the result of the author's original work.

## References


1. Sachs, H., Jeffrey, C., Kroll, G., Lafortune, G., Fuller, C., Woelm, F., *Sustainable Development Report 2022*, CUP, UK, 2022.
2. Tahat, M., Alananbeh, K., Othman, Y., Leskovar, D., Soil health and sustainable agriculture. *Sust.*, 12, 4859, 2020.
3. Wen, Y., Yao, W., Yu, T., Cheng, L., Zhang, Q., Yang, J., Zang, H., Long-term organic farming improves the red soil quality and microbial diversity in subtropics. *Agric. Ecosyst. Environ.*, 381, 109410, 2025.
4. Amarasinghe, A., Chen, C., Van Zwieten, L., Rashti, M.R., The role of edaphic variables and management practices in regulating soil microbial resilience to drought-a meta-analysis. *Sci. Total Environ.*, 912, 169544, 2024.
5. Shu, X., He, J., Zhou, Z., Xia, L., Hu, Y., Zhang, Y., Wang, C., Organic amendments enhance soil microbial diversity, microbial functionality and crop yields: A meta-analysis. *Sci. Total Environ.*, 829, 154627, 2022.
6. Diatta, A.A., Kanfany, G., Camara, B., Bassène, C., Manga, A.G., Seleiman, M., Schillaci, C., Compost as an alternative to inorganic fertilizers in cowpea [Vigna unguiculata (L.) Walp.] Production. *Legum. Sci.*, 6, e247, 2024.
7. Manzoor, A., Naveed, M.S., Ali, R.M.A., Naseer, M.A., Ul-Hussan, M., Saqib, M., Farooq, M., Vermicompost: A potential organic fertilizer for sustainable vegetable cultivation. *Sci. Hortic.*, 336, 113443, 2024.
8. Zhang, J., He, W., Wei, Z., Chen, Y., Gao, W., Integrating green manure and fertilizer reduction strategies to enhance soil carbon sequestration and crop yield: Evidence from a two-season pot experiment. *Front. Sustain. Food Syst.*, 8, 1514409, 2025.
9. Wu, Q., Zhang, J., Liu, X., Chang, T., Wang, Q., Shaghaleh, H., Hamoud, Y.A., Effects of biochar and vermicompost on microorganisms and enzymatic activities in greenhouse soil. *Front. Environ. Sci.*, 10, 1060277, 2023.

10. Oishy, M.N., Shemonty, N.A., Fatema, S.I., Mahbub, S., Mim, E.L., Raisa, M.B.H., Anik, A.H., Unravelling the effects of climate change on the soil-plant-atmosphere interactions: A critical review. *Soil Environ. Health,* 100130, 2025.
11. Huo, Y., Dijkstra, F.A., Possell, M., Singh, B., Ecotoxicological effects of plastics on plants, soil fauna and microorganisms: a meta-analysis. *Environ. Pollut.*, 310, 119892, 2022.
12. Zhang, J., Ren, S., Xu, W., Liang, C., Li, J., Zhang, H., Wang, K., Effects of plastic residues and microplastics on soil ecosystems: A global meta-analysis. *J. Hazard. Mater.*, 435, 129065, 2022.
13. Zhang, X., Yang, P., Lu, B., Artificial intelligence in soil management: The new frontier of smart agriculture. *Adv. Resour. Res.*, 4, 231, 2024.
14. McElhinney, J.M., Catacutan, M.K., Mawart, A., Hasan, A., Dias, J., Interfacing machine learning and microbial omics: a promising means to address environmental challenges. *Front. Microbiol.*, 13, 851450, 2022.
15. Basu, A. and Narayan, A., The role of machine learning in transforming agricultural practices: Insights into crop yield optimization and disease detection. *Iran J.Comput. Sci.*, 8, 1, 2025.
16. Aghababaei, A., Aghababaei, F., Pignitter, M., Hadidi, M., Artificial Intelligence in Agro-Food Systems: From Farm to Fork. *Foods*, 14, 411, 2025.
17. Akhter, R. and Sofi, S.A., Precision agriculture using IoT data analytics and machine learning. *J. King Saud Univ. Comput. Inf. Sci.*, 34, 5602, 2022.
18. Pace, R., Schiano Di Cola, V., Monti, M.M., Affinito, A., Cuomo, S., Loreto, F., Ruocco, M., Artificial intelligence in soil microbiome analysis: a potential application in predicting and enhancing soil health - a review. *Discov. Appl. Sci.*, 7, 85, 2025.
19. Pham, H.T., Awange, J., Kuhn, M., Evaluation of three feature dimension reduction techniques for machine learning-based crop yield prediction models. *Sens.*, 22, 6609, 2022.
20. Chang, A., Tang, T.M., Zikry, T.M., Allen, G.I., Unsupervised Machine Learning for Scientific Discovery: Workflow and Best Practices, *arXiv.2506.04553*, 2025.
21. Chikene, H., Singisetti, V., Pradhan, S.K., Revolutionising Agriculture with Augmented Reality (AR) and Virtual Reality (VR). *NG AI*, 1, 44, 2025.
22. Khosravi, M., Jiang, Z., Waite, J.R., Jonesc, S., Torres, H., Singh, A., Ganapathy subramanian, B., Singh, A.K., Sarkar, S., Optimizing Navigation and Chemical Application in Precision Agriculture with Deep Reinforcement Learning and Conditional Action Tree, *arXiv.2503.17985*, 2025.
23. Kastner, P., A Decoupled Framework for Fast and High-resolution Simulations of Annual Outdoor Thermal Comfort, Cornell Univ, 2022.
24. Wu, J., Tao, R., Zhao, P., Martin, N.F., Hovakimyan, N., Optimizing nitrogen management with deep reinforcement learning and crop simulations, in: *Proc. IEEE Comput. Vis. Pattern Recognit*, p. 1712, 2022.

25. Kumari, S., Raghuram, P., Venkatesh, V.G., Shi, Y., Future perspectives on progressive farming with adoption of virtual reality technology for sustainable quality in agriculture. *TQM J.*, 34, 250, 2022.
26. Wang, D., Cao, W., Zhang, F., Li, Z., Xu, S., Wu, X., A review of deep learning in multiscale agricultural sensing. *Remote Sens.*, 14, 559, 2022.
27. Alatawi, H., Albalawi, N., Shahata, G., Aljohani, K., Alhakamy, A.A., Tuceryan, M., Augmented reality-assisted deep reinforcement learning-based model towards industrial training and maintenance for nanodrop spectrophotometer. *Sens.*, 23, 6024, 2023.
28. Zhao, J., Fan, S., Zhang, B., Wang, A., Zhang, L., Zhu, Q., Research Status and Development Trends of Deep Reinforcement Learning in the Intelligent Transformation of Agricultural Machinery. *Agric.*, 15, 1223, 2025.
29. Pandey, K. and Saharan, B.S., Soil microbiomes: a promising strategy for boosting crop yield and advancing sustainable agriculture. *Discov. Agric.*, 3, 54, 2025.
30. da Silva, A.F., Ohta, R.L., Azpiroz, J.T., Fereira, M.E., Marçal, D.V., Botelho, A., Steiner, M., Artificial intelligence enables mobile soil analysis for sustainable agriculture, arXiv.2207.10537, 2022.
31. Bigonah, M., Jamshidi, F., Marghitu, D., Immersive agricultural education: Gamifying learning with augmented reality and virtual reality, in: *Cases on Collaborative Experiential Ecological Literacy for Education*, vol. 26, IGI Global, USA, 2024.
32. Whig, A., Whig, P., Gupta, S.K., Sustainability: Innovations and Impacts, in: *Exploring the Impact of Extended Reality (XR) Technologies on Promoting Environmental Sustainability*, vol. 38, p. 1, Springer, New York, 2025.
33. Marzouk, S.H., Kwaslema, D.R., Omar, M.M., Mohamed, S.H., Harnessing the power of soil microbes: Their dual impact in integrated nutrient management and mediating climate stress for sustainable rice crop production. A systematic review. *Heliyon*, 11, 2025.

# 6

# Intelligent Farm: An Automated Farming Technology Deploying Reinforcement Learning for Agroforestry Conservation Agriculture

**K. Kalaivanan[1]*, V. Bhanumathi[2] and Prasanth Aruchamy[3]**

*[1]School of Electronics Engineering, Vellore Institute of Technology, Chennai, India*
*[2]Department of Electronics and Communication Engineering, CEG Campus, Anna University, Chennai, India*
*[3]Department of Computer Science and Engineering, Vel Tech Rangarajan Dr. Sagunthala R&D Institute of Science and Technology, Chennai, India*

***Abstract***

Cloud computing and artificial intelligence (AI) are emerging technologies in all real-time applications. The automation of detecting and predicting water irrigation, pest control, and fertilizer usage is still crucial in agroforestry. The Internet of Things (IoT) is the most promising tool in implementing intelligent farming and improving land management. The goal of agroforestry is to preserve the environment and natural resources. Reinforcement learning (RL) is used for prediction and classification in smart agriculture; it considers how crop growth, yield, soil, environmental characteristics, climate, and watering of agricultural fields can change over time and space. The implementation of RL in agroforestry has potential in several real-time smart agricultural applications. This chapter explores the implementation of IoT-based intelligent farming, focusing on various decision-making systems. The cloud computing system in IoT provides extensive services in a centralized manner by sharing computation mechanisms, memory, and costs. Furthermore, it examines the requirement of placing computing techniques near end devices due to the rapid growth of IoT devices in real-time applications. It examines various AI and reinforcement learning techniques utilized to facilitate immediate smart decisions. This chapter also discusses the limitations, challenges,

**Corresponding author*: kalaivaanankk@yahoo.com

Anitha Velu, Prasanth Aruchamy, Raghu Ramamoorthy, Rajesh Kumar Dhanaraj and Seifedine Kadry (eds.) Climate Smart Agriculture, (101–124) © 2026 Scrivener Publishing LLC

applications, and future prospects of deploying AI and reinforcement learning in intelligent farming.

***Keywords:*** Artificial intelligence, automation, cloud computing, deep reinforcement learning, edge computing, Internet of Things, smart agriculture

## 6.1 Introduction to the Components of Intelligent Farming

Agroforestry conservation offers social, economic, and environmental advantages by combining trees and shrubs with crops and/or cattle on the same plot of land. Unpredictable environmental conditions and rapid increase in population growth create significant challenges for food production. Also, resource constraints and infeasibility conditions for the machinery cause economic losses. To overcome these issues, the researchers are looking for innovative solutions by applying advanced technology and obtaining high profitability in agricultural products. The sustainability of agricultural practices needs multi-criteria assessments in fire detection, water irrigation, soil condition, pesticide requirement, fertilizer utilization, and ripe harvesting period. A modernized process has evolved in agriculture for monitoring and managing farmland. Wireless sensor networks with IoT collect farmland information from sensors. Various computing servers, such as edge, fog, and cloud servers, are employed depending on the applications to process collected data and make decisions using artificial intelligence (AI). Reinforcement learning makes decisions based on the agent's interaction with the environment to maximize the reward function. Recent studies integrate AI and deep reinforcement learning (DRL) with IoT to automate and optimize decision-making for these complex agricultural challenges. It works especially well for complex, unpredictable, and dynamic decision-making processes like those found in agroforestry systems. Automated decision-making systems that support the management of different land use issues can be created using reinforcement learning, e.g., planting plans, crop rotation, pest control, irrigation, tree-planting schedules, and biodiversity preservation [1]. In recent years, most researchers have focused on smart farming to promote the agricultural process and increase productivity.

## 6.2 Big Data Analysis

The life cycle of big data guides authorities to understand and develop each stage of a smart system. The IoT devices generate enormous amounts of data and store them in the cloud server, which is analyzed and processed by the cloud computing paradigm, which performs four major processes: (1) data acquisition; (2) pre-processing; (3) analysis; and (4) decision-making and visualization [2]. The process of big data analysis is shown in Figure 6.1.

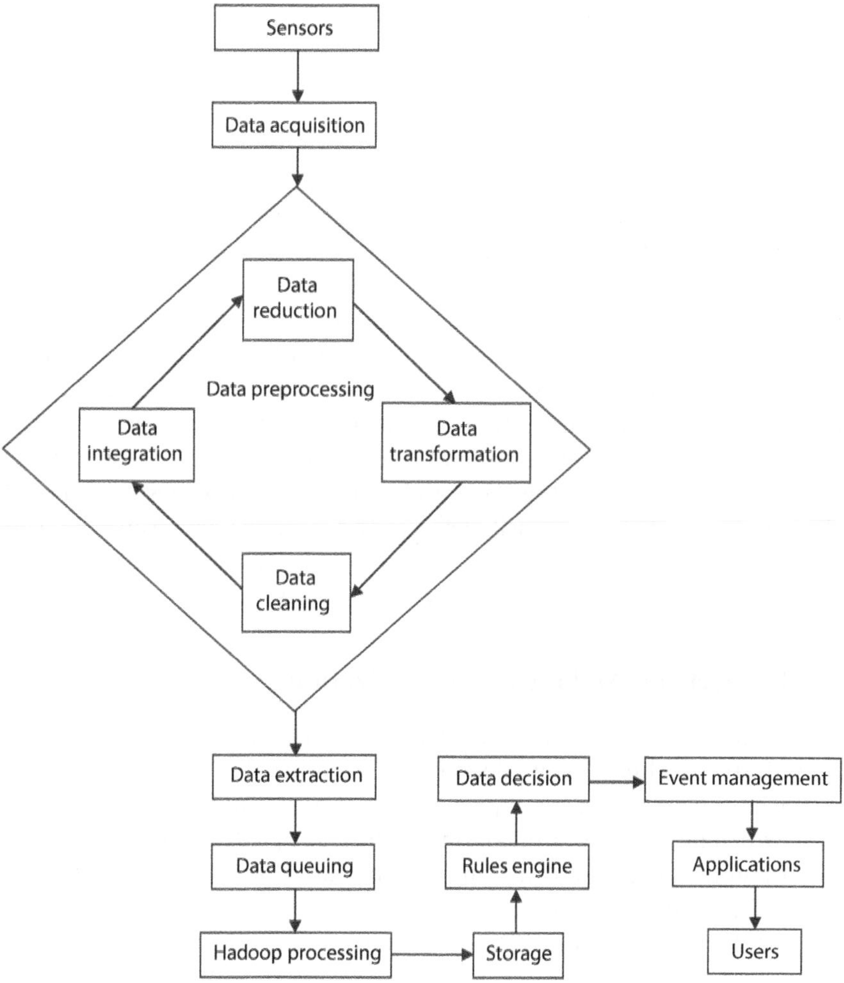


**Figure 6.1** Process of big data analysis.

### 6.2.1 Data Acquisition

Smart application developments are made up of various heterogeneous sensor nodes deployed in the targeted sensing area to collect information. These sensor nodes continuously obtain and transmit data to the cloud server.

### 6.2.2 Pre-Processing

The system removes redundant data, noise, and other errors from the collected data before sending the collected information to data processing. Thus, it minimizes the network traffic, processing time, and memory requirement.

### 6.2.3 Data Processing and Analytics

The data processing performs two major operations: (1) filtration and (2) data classification. Filtration is the process of removing unnecessary metadata and redundant data attached to the sensor data. This will easily perform the classifications depending on the data type and identifier. Data pre-processing is performed through the Spark and Hadoop ecosystems, which effectively process vast amounts of data at high speed through parallel systems. Hadoop is associated with the Hadoop distributed file system, which contains many slave nodes with at least one master node. The master node splits the large amount of data into the fixed size of chunks and distributes them to the slave nodes, as well as links results from all slave nodes, so that it can easily process and store the data using the map function. Spark is a real-time data processing engine that is combined with Hadoop to extract and process data effectively [3].

### 6.2.4 Decision-Making and Visualization

Finally, it performs decision-making and infers the results based on the optimization techniques and decision model utilized in the system design. Artificial intelligence is an important technique to build a smart application, which performs four major operations: (1) updating the performance of an automated decision making system based on the current requirement of the people by collecting the fact and experience of the surrounding area; (2) Replacing the physical function to ensure the autonomous system under unmanned conditions; (3) AI optimizing the efficiency by the experiences, volume of data and computing paradigm; and (4) deriving the efficient rules for automation based on the experience and facts.

## 6.3 Reinforcement Learning

Reinforcement learning is a subgroup of machine learning (ML) that contains neural networks (NN) and Q-learning. Reinforcement learning is a well-organized and unbiased optimization technique compared to supervised machine learning methods because it does not need a large amount of labeled datasets. It is modeled using a Markov decision process through which the agent learns how to act depending on the environment to update the reward function value. Due to a lack of matrix representation parameters for the large set of states and actions and insufficient memory space, RL is suitable for low-dimensional datasets, so it has a limitation in many real-time applications. Deep reinforcement learning uses NN to approximate the reward function and policy value for high-dimensional input data, whereas deep Q-learning (DQN) represents a vast set of states and action space [4]. The RL concept is shown in Figure 6.2.

Reinforcement learning uses a feedback mechanism applied between the agent (decision maker) and the environment to produce a sequence of decisions for uncertain conditions. The decision maker determines and decides based on the set of rules and environmental observations. The agent's learning process depends on its observation of the environment. At each iteration, it decides and controls action depending on the set of rules. Decision-making rules are adjusted through the reward value. The reward is a scalar value obtained by its experience and indicates the closeness of the decision taken by the agent. The important function of the decision-maker is to create an objective function based on the reward obtained at each iteration of the decision-making. This enhances the utility function of the reward values. This RL iteration is continuously repeated until the decision sequence finally ends. There are two types of RL: (1) centralized RL, and (2) individual RL. A centralized network connects all the systems, in which each system shares its experience with the network. This can help the agent make an optimized decision for the real-time application. In individual RL, each system maintains its own networks and decides based on its own experiences.

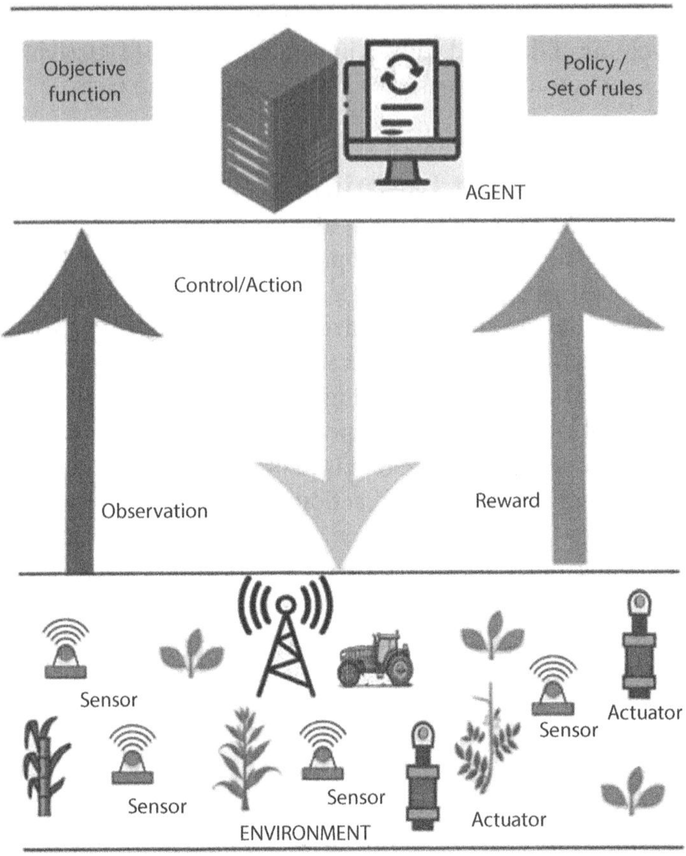


**Figure 6.2** The concept of the RL technique.

### 6.3.1 Markov Decision Process

The Markov chain model is a foundation for deriving a sequence of decisions in DRL. The agent updates the policy and obtains the best policy ($\varphi$) depending on the cumulative reward function [4]. The stochastic Markov decision process is described in Equation 6.1.

$$\varphi = (s, A, P, \beta) \tag{6.1}$$

Where $s = s_1, s_2 \ldots s_t \ldots s_n$ is the collection of all possible states of the system, $s_t$ is the state of the system at time $t, s_t \in s$, $A = a_1, a_2, \ldots a_n$ is the set of all possible actions, and $\beta$ is the reward value. The policy updates and obtains an action $a_t \in A$ by receiving a $s_t$. The probability (P) is calculated by the transition from $s_t$ to $s_{t+1}$. The next state $s_{t+1}$ is utilized to measure reward value $r_t$, $r_t \in \beta$. The reward function determines the current value of every future reward value by multiplying a discount factor $\gamma$.

The discount factor values lie between zero and one. The transition probability is calculated by using the Markov model as mentioned in Equation 6.2.

$$P = (s_{t+1} \mid s_t, a_t) \tag{6.2}$$

### 6.3.2 Q-Learning

Equation 6.3 is used to determine an optimal action cost.

$$Q(s, A) \leftarrow Q(s, A) + \tau[r + \beta Q(\hat{s}, A) - Q(s, A) \tag{6.3}$$

where $\tau$ is the learning rate.

The main drawback of this method is infeasible because of the large state space and large action space [5].

### 6.3.3 Deep Q-Learning

Q-learning is ineffective in high-dimensional state spaces, requiring a huge matrix to update the Q-value of every action based on the state of the Q-table. Also, the Q-learning algorithm may fail under this condition due to insufficient memory space. To overcome this drawback, DQN combines traditional Q-learning and deep NNs by measuring the Q-value based on experience, not on the environment. Deep Q-learning depends on a single parameter $\theta'$ to predict the target value and take an action [4], as mentioned in equation (6.4).

$$q = r + \beta max Q_\rho(\hat{s}\ \hat{a}\ \theta') \tag{6.4}$$

With its great fitting capabilities in image recognition, audio and video processing, text processing, and other areas, DQN is an effective tool for handling high-dimensional data.

### 6.3.4 Double Deep Q-Networks

The overestimation of Q-value leads to errors in the target value due to noise, non-stationary function, inadequate function approximation, among others. To overcome this issue, double DQN was introduced to represent the single value function into two networks [6]. The primary network determines the optimal actions, and the target network gives a value function that removes the need for an extra network. The primary network calculates the target value based on the parameter $\theta'$ as given in Equation 6.5.

$$q_i = r_{i+1} + \beta Q(s_{i+1}, argmax\ Q(s_{i+1}, a; \theta), \hat{\theta} \tag{6.5}$$

### 6.3.5 Dueling Deep Q-Networks

Dueling DQN was presented to improve policy determination for uncertain conditions, stability, and learning efficiency. In dueling DQN, CNN is used to extract the feature of the state and has two output streams: (1) state value function and (2) advantage function [7, 8].

The advantage function calculates the relative advantage of each action in a particular state as given in Equation 6.6.

$$\alpha(s,a) = q(s,a) - v(s,a) \tag{6.6}$$

The specified state's value is provided by the state value function, irrespective of the action performed.

Equation 6.7 is used to measure the Q value of each action $q(s,a)$.

$$q(s,a;\theta,\xi,\varpi) = v(s;\theta,\varpi) + \alpha(s,a;\theta,\xi) \tag{6.7}$$

where $\xi$ and $\varpi$ are the parameters of fully connected layers from the two output streams.

The main problem in DRL was instability during training due to handling a huge set of actions and state space, leading to divergence in the training process. To overcome this, divergence parameters removal was performed by the agent through its experience (experience replay) during the learning process. Also, the Q-Learning-based update process utilized separate networks to generate the target. The training process was performed by an RNN and DQN agent, in which the RNN was applied to pre-train the data, and the agent took actions based on the greedy policy [9].

Distributional reinforcement learning deployed the distribution instead of an expected value to model cumulative rewards. The existing RL utilized an expected value to obtain cumulative rewards, which fails to provide detailed information about the uncertainty of rewards as given in Equation 6.8. Distributional RL gives more comprehensive reward information, denoted by $Z(s, A)$ which helps improve decision accuracy.

$$Q\left(s_t, A_t\right) = E[Z(s_t, A_t)] = E\left(\sum_{i=0}^{i=\infty} \gamma_{t+i} R(s_{t+i}, A_{t+i})\right) \tag{6.8}$$

The success rate of each iteration was predicted using probability distribution and Bayes' theorem to maximize the reward function [10]. It is calculated by using Equation 6.9.

$$P\left(\frac{\eta}{\mu}\right) = \frac{P\left(\frac{\mu}{\eta}\right) P(\eta)}{P(\mu)} \tag{6.9}$$

$\mu$ is the observation, $P\left(\frac{\eta}{\mu}\right)$ is the probability of success rate given observation, $P\left(\frac{\mu}{\eta}\right)$ is the probability of the observation given the success rate, $P(\eta)$ is the probability distribution of the success rate, $P(\mu)$ is the probability of success and penalty.

## 6.4 Need for the Internet of Things in Smart Applications

The IoT integrates several technologies, including wireless sensor networks, RFID, cloud server, computing paradigms, communication systems, and end-user applications. Innovation in wireless communication and sensor technologies is mainly utilized in the farm to collect crop land information. Various ML algorithms are then deployed to analyze and make effective decisions [11]. This capability has made it a good choice for

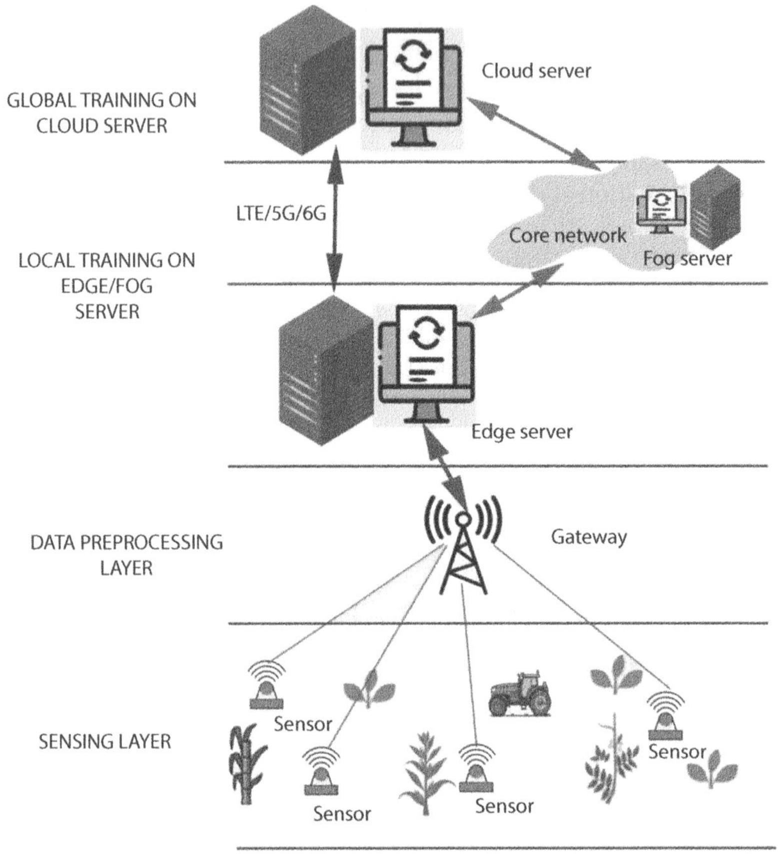


**Figure 6.3** Smart agriculture with IoT components.

environmental monitoring, process control automation, and traceability in agriculture and the food industry. The IoT provides computing tools, decision-making algorithms, software, and hardware platforms that seamlessly integrate knowledge, services, and resources for ensuring a better economy in terms of quality, productivity, and profit [12]. Smart agriculture with IoT components is depicted in Figure 6.3.

## 6.4.1 Function of Internet of Things Elements

### *6.4.1.1 Cloud Computing*

The cloud server can be operated remotely thanks to innovations in communication technologies, and it is situated far from the data source. It provides sophisticated computing, processing, and storage resources for efficient data analysis. The majority of computations are managed and controlled centrally in cloud computing [13]. This computing paradigm experiences longer delays and gets poor user feedback in real-time applications. For example, nowadays various types of surveillance cameras are utilized in the smart city development; due to longer delays and security concerns, cloud-computing-based video analytics are not suitable for this type of application.

#### *6.4.1.2 Fog Computing*

Fog computing is an innovative technique situated between sensors and cloud computing. It is used to overcome the shortcomings (i.e., latency in response time) that occur in the cloud during real-time applications. The fog node is placed near the edge of the things that extend their services to the cloud. The fog node accesses and processes real-time data collected from the IoT sensor and gives a real-time notification to users [15]. Fog nodes are placed near the edge of IoT devices with centralized cloud computing, which have high processing and storage capabilities [14]. For example, a user wants to review the IoT security camera video stream through the handheld device. Here, a fog node is responsible for providing a video stream to the user instead of cloud services, responding much quicker than cloud services. Fog computing was designed by Cisco in 2012. Fog computing processes information within the LAN on gateway nodes or fog nodes. Fog computing offers many advantages over cloud computing, including better performance, enhanced bandwidth utilization, high security and resiliency, enhanced support for mobile and location-aware services, and low latency in real-time applications, among others.

#### *6.4.1.3 Edge Computing*

The computation takes place directly on the processor attached to the sensor node or devices, so that it handles data without shifting the data to any place. In other words, the data is processed within the local edge IoT devices, thus the overall real-time performance is highly improved [16]. For example, (1) a tablet or smartphone acts as the edge device between the human body and the cloud, (2) a gateway between the industrial things and the cloud, and (3) a cloudlet acts as the micro data center, which is the edge device between the mobile phone and the cloud [17]. In edge computing, things act as data customers and data generators. Edge computing

**Table 6.1** Performance analysis of various computing techniques.

| Performance | Cloud computing | Fog computing | Edge computing |
|---|---|---|---|
| Storage | High | Restricted | Restricted |
| Interactions in real time | Yes | Yes | Yes |
| Processing power | High | Restricted | Restricted |
| Storage | High | Restricted | Restricted |
| Accuracy | High | Restricted | Restricted |
| Real-time response | Delayed | Fast | Fast |
| Security issue | High | Low | Low |
| Latency issue | High | Low | Low |
| In-depth analysis | Highly possible | Possible | Restricted |
| Cost | Low | Moderate | High |
| Internet | Required constantly | May or May not | May or May not |
| Network traffic | High | Low | Low |
| Bottleneck problem | High | Low | Low |
| Response time | Few seconds | Milli seconds | Milli seconds |
| Higher-order computation | Highly possible | Limited | Limited |
| Location aware | No | Yes | Yes |
| Mobile support | No | Yes | Yes |
| Data access | Anywhere, regardless of physical location | Restricted | Restricted |

(*Continued*)

**Table 6.1** Performance analysis of various computing techniques. (*Continued*)

| Performance | Cloud computing | Fog computing | Edge computing |
|---|---|---|---|
| Operation | Centralized manner (i.e., operation performed only in a limited place) | Focus on the infrastructure side | Focus very closely to the things side |
| Communication between client and server | Multi-hop | Direct hop | Direct hop |
| Geographic distribution | Centralized | Distributed | Distributed |
| Server count | Few | Large | Large |

distributes the request and delivery service between the cloud and the user and also provides local storage, analytical tools, processing, and other network resources to the resource-constrained devices. The edge server pre-processes the data so that only a small amount of data is transferred to the central data center. Thus, it reduces the network load and congestion. For example, a manufacturing instrument may have a huge number of sensors used not only for monitoring but also for adjusting the manufacturing process based on locally collected IoT sensor information in real-time. Internet-of-Things devices can easily transfer their data to the edge server when the amount of data created exceeds the processing capacity, and also provide services close to the IoT devices, ensuring low latency. The comparison of various computing methods is given in Table 6.1.

## 6.5 Challenges of the Internet of Things

### 6.5.1 Scalability

Scalability is an important requirement in IoT in two aspects: (1) big data, and (2) geo-distribution. The volume of data gathering and processing increases when a huge number of end devices and computing devices are connected to the IoT, and the geographical distribution increases. First, the handling of big data is a major characteristic of the IoT to scale the processing of large data sizes. Edge and fog computing can handle distributed services such as computing, storage, processing, and applications in large geographic areas where the density of the nodes increases [18].

### 6.5.2 Interoperability

Interoperability is an important requirement of the IoT that connects to various heterogeneous platform environments. In a real scenario, different types of sensors and devices are deployed to gather data, having heterogeneity in data types, languages, message sizes, data rates, energy consumptions, etc. The heterogeneity of messages and devices requires syntax, semantics, and communication interface interoperability. Due to varying resource availability in heterogeneous devices, automated software is needed to handle and standardize the heterogeneous data. This is used to understand and share information between two different devices by creating a unified data stream, computing tools, and services [19].

### 6.5.3 Latency

The deployed sensor nodes generate a massive volume of data at high data rates from real-time applications and send it to the cloud server. The data may experience high latency and loss during a real interaction due to the dynamic nature of data traffic and rate in the IoT, and is exchanged with the cloud server. Data preprocessing removes unwanted information from the sensor's raw data, reducing the data traffic entering the networks and congestion. The deployment of fog and edge computing reduces the latency by placing the analytical tools and storage near the source of information. Fog nodes should be present at the exact location of the target area in real-time applications, enabling effective decision-making and real-time execution [20].

### 6.5.4 Security

Security is a crucial requirement of the computing paradigm in IoT. In cloud computing, numerous IoT devices are interconnected to cloud servers to share their information. For security concerns, fog and edge computing are highly preferable. Due to the growth of ubiquitous and pervasive systems, security becomes a crucial factor in protecting personal data [21].

### 6.5.5 Location Awareness

The IoT applications should know the location where the event is collected, enhancing the adaptability and accuracy of the event detection and response. The location-aware data collection can enhance IoT performance by providing a high degree of consciousness and resilience to the real-world scenario [22]. Thus, it has improved accuracy, execution, and quality of the deployed applications. Therefore, the deployed computing paradigm should know the location of sensed information and the node deployment to enhance the performance of adaptation modules and data processing.

### 6.5.6 Mobility

Mobility support is a significant challenge to the recent growing ubiquity of mobile devices in IoT-based real-time applications. The Internet of Medical Things should be capable of supporting the execution and run-time requirements of the dynamic environment. Also, computing paradigms should adjust themselves to manage high mobility and sustain their operation without any interruptions [23]. Fog computing is highly preferable for mobile support, in which fog nodes should adapt to the mobile environment and can move the fog node authority from one to another without causing problems.

### 6.5.7 Quality of Services

Quality of service (QoS) is an important requirement for many real-time applications. Real-time response is a highly dependent parameter of QoS measures in many emergency applications. The cloud-based system takes a long time to process data; it also faces high network traffic and congestion. To achieve effective QoS, edge and fog computing are introduced to collect and process the relevant multidimensional sensor data to complete specific tasks efficiently. It also ensures the accuracy of collected data and provides uninterrupted data services continuously [24].

### 6.5.8 Availability

The platform should support critical IoT application requirements and ensure system reliability. Cloud services such as software, analytical tools, skills, standards, processing capability, Internet, wireless coverage, security, and so on are always available to authorized users. The IoT services or devices are immediately delivered in a real-time fashion whenever it is requested by users. However, the denial of service (DoS) is the most serious threat to the availability of cloud services [25].

## 6.6 Smart Agriculture Applications

The combination of RL and IoT in smart agriculture results in a strong system for controlling agricultural activities in a more efficient, flexible, and sustainable manner. Real-time data may be collected from agricultural situations using IoT-enabled sensors and devices, giving RL algorithms the input they need.

Rewards in RL: This is nothing but the feedback an agent receives after completing an action is the "reward". The examples of reward evaluation are as follows:

Positive reward: Reduced insect damage, increased agricultural output, or effective use of resources (fertilizer, water).

Negative reward: Crop damage, decreased yield, or resource waste (such as excessive pesticide use or irrigation).

Sustainability reward: Advantages for the environment, such as less carbon emissions, water conservation, and chemical runoff.

An application of RL with IoT in smart agriculture is shown in Figure 6.4. A modified ant colony-enabled DQN model is presented to schedule tasks to a computing server and use RL to execute and forecast tasks for agricultural process monitoring [26]. The value assigned to the features of the farmer's goods and services is updated continuously by the reward learning algorithm. This can be used to decide the agricultural process [27]. Using a cloud server and DRL, an intelligent assessment is made instantly to predict the irrigation schedule in the agricultural process [28, 29]. Reinforcement learning, along with the Markov decision process, is deployed to predict the right amount of water required by the crop field [30]. Fuzzy logic was used to pre-process the raw data before it is sent to the DRL, which changes the policy during decision-making and forecasts the probability values of each action [31]. The online RL-based method lowers the expense of edge computing, which relies on local data for job scheduling and dispatching decisions [32]. Reinforcement learning-based energy-aware offloading and delay minimization in task processing was introduced in [33]. Environmental data was collected by deploying an IoT-enabled UAV on an agricultural farm. Based on the UAV's remaining energy, the offloading decision is made. The multi-agent-based RL was

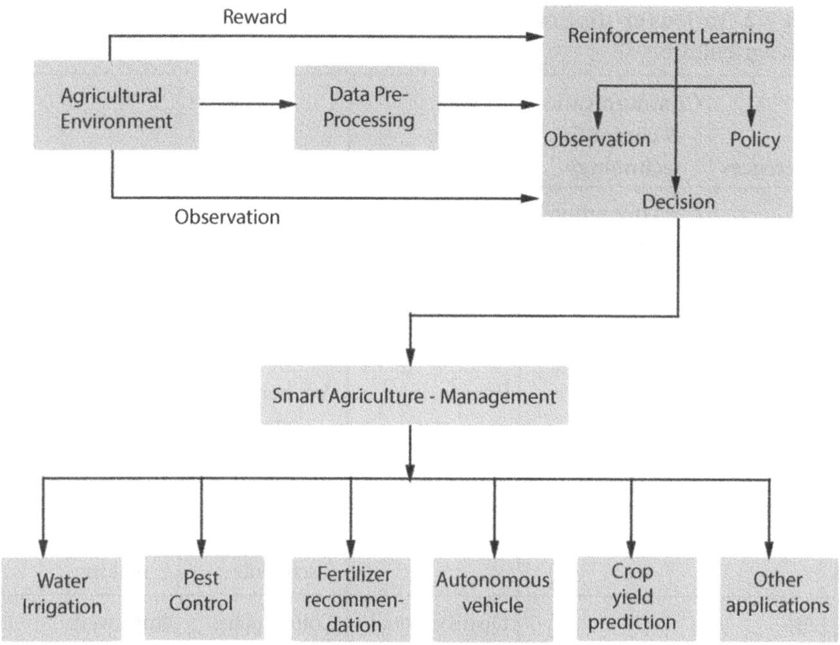


**Figure 6.4** Application of RL with IoT in smart agriculture.

introduced to make effective decisions in field coverage and increase the reliability for highly dynamic environment data [34]. Reinforcement learning was applied to predict crop rotation by using a simulated agricultural computing environment with economic strategies. The particle swarm optimization technique was applied for estimating the cost at harvest and additional subsidies near the initial value [35]. However, the experience replay stores the next transition state in the memory, which is highly correlated and non-stationary. This may lead to the convergence issue during the training. A DRL-based supply chain management system was proposed [36] to overcome the convergence issues and obtain a stable decision in the action of agri-food production and storage. A summary of various existing RL methodologies with IoT in smart agriculture, discussed in this chapter, has been illustrated in Table 6.2.

Challenges of implementing RL in intelligent farming:

- Because agroforestry systems are intricate, the RL system needs precise, up-to-date data to function properly. To gather valuable data, there must be sufficient sensors, IoT devices, and data infrastructure.

**Table 6.2** Various existing reinforcement learning methodologies with IoT in smart agriculture.

| References | Communication & computing technology | Methodology | Applications | Remarks |
|---|---|---|---|---|
| [26] | IoT, computing server | RL, DQN, ACO | Monitoring agricultural process | Prioritizing the task, reducing the latency, and energy consumption |
| [27] | - | RL | Post harvests – offers | Every decision supports the actual choices. |
| [28] | IoT, cloud server | RL | Water irrigation schedule | Instant smart decision making |
| [29] | - | Distributional RL | Cotton plant – irrigation and crop yield | Improved accuracy in decision-making over traditional RL |
| [30] | IoT | Deep RL | Smart irrigation management | - |
| [31] | IoT, edge server | Fuzzy-based Deep RL | Frost forecast | Increased complexity |
| [32] | IoT, edge computing | DRL | Tomato plant disease | Does not require global information. |
| [33] | IoT, edge computing, UAV | RL | Energy prediction of UAV and decision on the smart farm process | Reduced the delay in the decision process |

*(Continued)*

**Table 6.2** Various existing reinforcement learning methodologies with IoT in smart agriculture. (*Continued*)

| References | Communication & computing technology | Methodology | Applications | Remarks |
|---|---|---|---|---|
| [10] | - | RL, Probability distribution, and Bayes' rule | Optimal placement of a sensor in an agricultural field | Reduced exploration |
| [34] | IoT, UAV | RL | Agricultural field coverage by the UAV | Improved decision-making and reliability in a dynamic environment |
| [35] | - | RL, particle swarm optimization, Long short-term memory | Cost and subsidy prediction for crop planning and agricultural products | Improved computational method for the agriculture field and crop rotation |
| [9] | - | RNN, DQN | Crop yield prediction | Improved stability in DRL's training process |
| [36] | IoT | DRL, Blockchain, random sampling in experience replay | Agri-food supply chain | Stable and easier to converge in decision-making |

- Because agroforestry involves so many different variables and processes, it is challenging to develop models that fully capture its complexity. This complexity must be included in the design of the RL algorithm.
- To train, RL systems frequently need a large amount of data and time. This could imply that it may take multiple growing seasons for the system to function at its best in an actual agroforestry setting.
- Human expertise is still crucial, even though RL systems can automate a large portion of the decision-making process. The expertise of local farmers on climate, species interactions, and land management will remain important.

## 6.7 Conclusion

In this chapter, a DRL-based intelligent system was studied to make smart decisions in improving crop growth by considering crop land factors. Owing to a number of problems in agroforestry, researchers are motivated to provide a novel way to boost agricultural output and protect the environment and croplands. Deep reinforcement learning is one of the approaches being studied, which has demonstrated effective decision-making in the complex concerns of farmland management. There is enormous potential to increase production, maximize the use of natural resources, and improve environmental sustainability by combining reinforcement learning with agroforestry conservation. Reinforcement learning-based technologies are an important tool for contributing to the long-term viability of agroforestry systems by utilizing real-time data, adaptive decision-making, and continuous learning. This will ultimately help to achieve global conservation goals and guarantee food security for future generations. This chapter summarizes the various existing DRL methods utilized in smart farming applications, particularly crop growth, environmental factor monitoring, and water irrigation. The integration of DRL with IoT in intelligent farming gives promising prospects in promoting agricultural automation and improving agricultural productivity. This chapter demonstrates the wide range of assistance that will be available in the future with 5G/6G-enabled smart agriculture, offering continuous connectivity services for the rapidly growing IoT and end devices.

**AI disclosure statement:** The author confirms that no artificial intelligence (AI)-generated content or AI-assisted writing tools were used in the preparation of this chapter (including summaries and the abstract). All text, analysis, and conclusions are the result of the author's original work.

## References


1. Jha, K., Doshi, A., Patel, P., Shah, M., A comprehensive review on automation in agriculture using artificial intelligence. *Artif. Intell. Agric.*, 2, 12, 2019.
2. Pulikottil, T., Estrada-Jimenez, L.A., Abadia, J.J.P., Carrera-Rivera, A., Torayev, A., Rehman, H.U., Big Data Life Cycle in Shop-Floor–Trends and Challenges. *IEEE*, 11, 30008, 2023.
3. Geng, D., Zhang, C., Xia, C., Xia, X., Liu, Q., Fu, X., Big Data-Based Improved Data Acquisition and Storage System for Designing Industrial Data Platform. *IEEE*, 7, 44574, 2019.
4. Din, A., Ismail, M.Y., Shah, B., Babar, M., Ali, F., Baig, S.U., A deep reinforcement learning-based multi-agent area coverage control for smart agriculture. *Comput. Electr. Eng.*, 101, 108089, 2022.
5. Ali, N., Wahid, A., Shaw, R., Manson, K., A reinforcement learning approach to dairy farm battery management using Q learning. *J. Energy Storage*, 93, 112031, 2024.
6. Iqbal, A., Tham, M., Chang, Y.C., Double Deep Q-Network-Based Energy-Efficient Resource Allocation in Cloud Radio Access Network. *IEEE*, 9, 20440, 2021.
7. Wu, X., Chi, J., Jin, X., Deng, C., Reinforcement learning approach to the control of heavy material handling manipulators for agricultural robots. *Comput. Electr. Eng.*, 104, 108433, 2022.
8. Ban, T., An Autonomous Transmission Scheme Using Dueling DQN for D2D Communication Networks. *IEEE Veh. Technol. Mag.*, 69, 16348, 2020.
9. Elavarasan, D. and Vincent, P.M.D., Crop Yield Prediction Using DRL Model for Sustainable Agrarian Applications. *IEEE*, 8, 86886, 2020.
10. Uyeh, D.D., Bassey, B.I., Mallipeddi, R., Asem-Hiablie, S., Amaizu, M., Woo, S., Ha, Y., Park, T., A Reinforcement learning approach for optimal placement of sensors in protected cultivation systems. *IEEE*, 9, 100781, 2021.
11. Sethi, P. and Sarangi, S.R., Internet of Things: Architectures, Protocols, and Applications. *J. Electr. Comput. Eng.*, 17, 9324035, 2017.
12. Ren, J., Yu, G., He, Y., Li, G.Y., Collaborative Cloud and Edge Computing for Latency Minimization. *IEEE Veh. Technol. Mag.*, 68, 5031, 2019.
13. Donno, M.D., Tange, K., Dragoni, N., Foundations and Evolution of Modern Computing Paradigms: Cloud, IoT, Edge, and Fog. *IEEE*, 7, 150936, 2019.
14. Cha, H., Yang, H., Song, Y., A Study on the Design of Fog Computing Architecture Using Sensor Networks. *Sensors*, 18, 3633, 2018.

15. Chiang, M. and Zhang, T., Fog and IoT: An Overview of Research Opportunities. *IEEE Internet Things J.*, 3, 854, 2016.
16. Fan, Q. and Ansari, N., Application Aware Workload Allocation for Edge Computing based IoT. *IEEE Internet Things J.*, 5, 2146, 2013.
17. Pan, J. and McElhannon, J., Future Edge Cloud and Edge Computing for Internet of Things Applications, Future Edge Cloud and Edge Computing for Internet of Things Applications. *IEEE Internet Things J.*, 5, 439, 2018.
18. Achar, S., Neural-Hill: A Novel Algorithm for Efficient Scheduling IoT-Cloud Resource to Maintain Scalability. *IEEE*, 11, 26502, 2023.
19. Goudari, M., Wu, H., Palaniswami, M., Buyya, R., An Application Placement Technique for Concurrent IoT Applications in Edge and Fog Computing Environments. *IEEE Trans. Mob.*, 20, 1298, 2021.
20. Nassar, A. and Yilmaz, Y., Reinforcement Learning for Adaptive Resource Allocation in Fog RAN for IoT With Heterogeneous Latency Requirements. *IEEE*, 7, 128014, 2019.
21. Shirazi, S.N., Gouglidis, A., Farshad, A., Hutchison, D., The Extended Cloud: Review and Analysis of Mobile Edge Computing and Fog From a Security and Resilience Perspective. *IEEE J. Sel. Areas Commun.*, 35, 2586, 2017.
22. Perera, C., Talagala, D.S., Liu, C.H., Estrella, J.C., Energy-Efficient Location and Activity-Aware On-Demand Mobile Distributed Sensing Platform for Sensing as a Service in IoT Clouds. *IEEE Trans. Comput. Soc. Syst.*, 2, 171, 2015.
23. Qadir, J., Sainz-De-Abajo, B., Khan, A., Garcia-Zapirain, B., Torre-Diez, I.D.L., Mahmood, H., Towards Mobile Edge Computing: Taxonomy, Challenges, Applications and Future Realms. *IEEE*, 8, 189129, 2020.
24. Herrera, J.L., Galan-Jimenez, J., Foschini, L., Bellavista, P., Berrocal, J., Murillo, J.M., QoS-Aware Fog Node Placement for Intensive IoT Applications in SDN-Fog Scenarios. *IEEE Internet Things J.*, 9, 13725, 2022.
25. Ghahramani, M., Javidan, R., Shojafar, M., Taheri, R., Alazab, M., Tafazolli, R., RSS: An Energy-Efficient Approach for Securing IoT Service Protocols Against the DoS Attack. *IEEE Internet Things J.*, 8, 3619, 2021.
26. Devarajan, G.G., Nagarajan, S.M., Ramana, T.V., Vignesh, T., Ghosh, U., Alnumay, W., DDNSAS: Deep reinforcement learning based deep Q-learning network for smart agriculture system. *Sustain. Comput.: Inform. Syst.*, 39, 100890, 2023.
27. Lermen, F.H., Martins, V.L.M., Echeveste, M.E., Ribeiro, F., Peralta, C.B.L., Ribeiro, J.L.D., Reinforcement Learning system to capture value from Brazilian post-harvest offers. *Inf Process Agric.*, 11, 499, 2024.
28. Bu, F. and Wang, X., A smart agriculture IoT system based on deep reinforcement learning. *Future Gener. Comput. Syst.*, 99, 500, 2019.
29. Chen, Y., Lin, M., Yu, Z., Sun, W., Fu, W., He, L., Enhancing cotton irrigation with distributional actor–critic reinforcement learning. *Agric. Water Manag.*, 307, 109194, 2025.

30. Campoverde, L.M.S., Tropea, M., Rango, F.D., *An IoT based Smart Irrigation* Management System using Reinforcement Learning modeled through aMarkov Decision Process. *IEEE*, 12, 234, 2021.
31. Herabad, M.G. and Afshar, N.P., Fuzzy-based Deep Reinforcement Learning for Frost Forecasting in IoT Edge-enabled Agriculture. *IEEE*, 8, 132, 2022.
32. Yang, Z., Chiu, T., Sheu, J., Reinforcement Learning-Based Task Offloading of MEC-Assisted UAVs in Precision Agriculture. *IEEE*, 7, 183, 2023.
33. Nguyen, A.C., Pamuklu, T., Syed, S., Kennedy, W.S., Erol-Kantarci, M., Reinforcement Learning-Based Deadline and Battery-Aware Offloading in Smart Farm IoT-UAV Networks. *IEEE*, 8, 109, 2022.
34. Marwah, N., Singh, V.K., Kashyap, G.S., Wazir, S., An analysis of the robustness of UAV agriculture field coverage using multi-agent reinforcement learning. *Int. J. Inf. Technol.*, 15, 2317, 2023.
35. Wang, X., Xu, Y., Wang, H., Kang, M., Hua, J., Wang, F., Region-Farm crop planning through double deep Q-Learning toward sustainable agriculture. *IEEE Trans. Comput. Soc. Syst.*, 11, 7608, 2024.
36. Chen, H., Chen, Z., Lin, F., Zhuang, P., Effective management for blockchain based agri-food supply chains using deep reinforcement learning. *IEEE*, 9, 36008, 2021.

7

# Overcoming Challenges of Data Privacy, Security, and Scalability for Commercial Grain Farming

**S. Venkatesh[1], D. Jeevitha[2*], B. Senthilkumaran[3] and K. K. Devi Sowndarya[4]**

*[1]Department of Data Science and Business Systems, Faculty of Engineering and Technology, SRM Institute of Science and Technology, Chennai, India*
*[2]Department of Computer Science and Engineering, Jeppiaar Engineering College, Chennai, India*
*[3]Department of Computer Science and Engineering, Vel Tech Rangarajan Dr. Sagunthala R&D Institute of Science and Technology, Chennai, India*
*[4]Department of Electronics and Communication Engineering, DMI College of Engineering, Chennai, India*


## Abstract

Climate-smart agriculture must be promoted for food security and environmental sustainability; however, its diffusion will be impeded by related data privacy, security, and scalability issues. This chapter discusses the challenges in the diffusion of data-intensive agricultural technologies: Privacy, security, and scalability issues. This chapter explores how federated learning, differential privacy, and blockchain can solve such problems. Federated learning enables a collaborative approach to train the model without transferring raw data, hence ensuring enhanced privacy and security. Differential privacy adds noise to the data to ensure privacy is well protected, and yet it preserves data utility. Blockchain provides an assured method for recording and sharing agricultural data safely and transparently. To further strengthen data protection, the need for farmer-centric data ownership policies must be highlighted. Such policies empower farmers to control their data and ensure fair compensation and privacy. International standards for agricultural data security are also necessary to safeguard sensitive information and facilitate cross-border data sharing. With these challenges in mind and the power of


**Corresponding author*: jeevithadamotharanjpr@gmail.com; ORCID: 0009-0008-3338-924X

Anitha Velu, Prasanth Aruchamy, Raghu Ramamoorthy, Rajesh Kumar Dhanaraj and Seifedine Kadry (eds.) Climate Smart Agriculture, (125–146) © 2026 Scrivener Publishing LLC

technology, the potential of data-driven agriculture can be unlocked to create a more sustainable future.

***Keywords*:** Agriculture data, data privacy, data security, data scalability, blockchain

# 7.1 Introduction

## 7.1.1 Application of Data in Agriculture

The rapid advancement of data and its tools in agriculture has given way to what is referred to as the age of "smarter" or "digital" agriculture. The increasing use of technology in farming is primarily attributed to the pressing need to feed the projected 9.7 billion population by 2050, which may face even more pronounced obstacles, such as climate change, depletion of natural resources, and the inability of conventional agriculture to suffice. Access to agricultural carry out of practical introduction (precision agriculture) performance indicators, in contrast to simple "what should be done" dysregulation, empowers farmers to leap over the frustrations of presumptive decision-making. A farmer can hydrate his plants with just the right amount of water by broadcasting information on how much moisture is present in the soil at a given point in time [1]. This is only one of the many applications that aim to enhance productivity through various data visualization methods.

This chapter begins with the diagnosis of agricultural data recorded using different technologies. Satellites and drones facilitate the 'big' data acquisition on crop and soil conditions, as well as crop protection data, such as insect and disease pest analysis. While parachuted sensors tell you the broader perspective, they count readings of temperature, humidity, and nutrient density close to the growing plants, which might be much more accurate than any items just listed. However, the most complex processes relating to data integration are when this wealth of data is being mined for insights. This is where rivals in the market for more sophisticated farm management software come in, aggregating and analyzing all the available information and providing farmers with this knowledge. It is a method known as blockchain technology.

Farmers can decide using the data available to them, which may include structuring farming activities themselves, various stages of crop cycles, and up to marketing strategy development [2]. However, this data-driven agricultural transformation is also associated with some challenges. There is a need to protect farmers' privacy and data ownership. These are important circumstances because they are necessary for avoiding data abuse and for allowing farmers to keep control of essential data. Also, to make this possible for the majority of people, especially in rural areas, adequate data facilities, such as a good internet connection and data security, are very important.

### 7.1.2 Data Challenges in Climate-Smart Agriculture

The growing climate crisis increasingly threatens global food security, with innovative solutions for sustainable, resilient agriculture. Climate-smart agriculture (CSA) is an emerging practice that holds great promise, aiming to increase agricultural productivity sustainably, help adapt to climate change, and reduce greenhouse gas emissions. The implementation of CSA relies heavily on maximizing data-based technological potential. Through advanced analytics technology, artificial intelligence, and machine learning algorithms, farmers obtain better resource management decisions, improving the total agricultural output. The widespread computerization of the farming industry requires solutions to maximize the benefits of data-based agricultural systems. The chapter identifies major issues pertaining to commercial grain farming, data privacy and security, and scalability concerns. The chapter examines both farmer-proprietorship data policies and established global guidelines for agricultural data protection. The necessary challenges will be addressed to build an innovative and ethical data-driven agricultural system.

## 7.2 Data Privacy in Agriculture

### 7.2.1 Data's Significance in Agriculture

The modern farming industry allows digital documentation of extensive data from cultivated crops, the environment, and machinery. The gathered information helps farmers make sound choices to boost their agricultural output and develop environmentally friendly practices [3]. Digital farming technologies operate effectively when farmers surrender their data to agribusiness companies, which transform it into analytical farming solutions.

Most farmers avoid sharing their farm details due to fear of privacy invasion and because trust remains low, so they fear their data will be manipulated improperly. Regular agricultural operations lack knowledge about the multitude of information companies extract from their operation. Farmers worry about unwelcome data misuses, which include market decisions made by unauthorized entities and unapproved third-party distribution of their information. The controversy about which party holds ownership rights to farm sensor information and equipment data involves a conflict between farmers and technical corporations. Conflicting ownership terms for data prevent the exchange of information because these terms remain undefined. The abilities of agribusiness firms to process agricultural data, along with their financial strength, surpass those of individual farmers who must provide their data to outside entities.

### 7.2.2 Initiatives to Mitigate Privacy Issues

The relationship between farmers and agribusinesses received a boost through multiple framework systems, which helped manage agricultural privacy. Among the programs established, there are procedures for sharing agricultural information to protect the rights of farmers. As base components, the EU Code of Conduct on Agricultural Data Sharing framework requires transparency and fairness for its establishment. While farm owners maintain sole data ownership rights, the system requires explicit consent from every owner for agribusiness access. The guidelines establish both security requirements for agricultural data protection and protocols

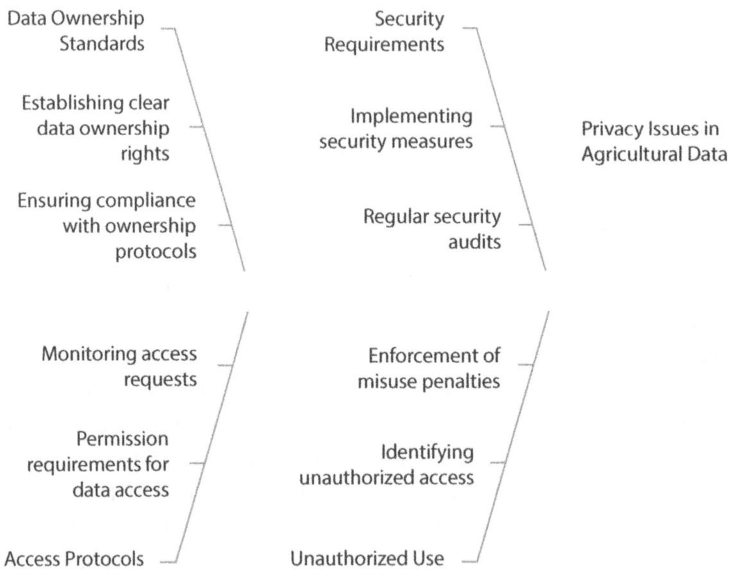


**Figure 7.1** Analyzing privacy issues in agricultural data.

defining authorized access and data ownership standards for preventing unauthorized use or misuse of information. Figure 7.1 demonstrates how privacy issues affect agricultural data.

The General Data Protection Regulation (GDPR) imposes strict rules on farmers regarding the collection then storing of data before processing. Every person, from farmers to ordinary citizens, possesses GDPR-defined rights for data control under its provisions. Implementation issues related to the GDPR emerge in agricultural data collection procedures because data monitoring stems from Internet of Things (IoT) sensors, along with drone operations and remote system uses. The GDPR law enforcement proves challenging for farmers because their data resides across numerous scattered locations, although many farmers remain unaware of the GDPR rights available to them. The GDPR offers vital legal guarantees for farm data, although existing implementation challenges need to be resolved.

The use of ethical contracts has proven to be an essential mechanism to address privacy-related issues. The mandatory informed consent agreements provide farmers with details about data usage before they share information. Photo-sharing agreements require complete details on how data can be used, along with security measures to establish mutual trust among participants. The combination of federated learning technology with differential privacy serves as a privacy-enabling tool that safeguards agricultural databases. Establishing legal structures and ethical agreements utilizing technological security protocols enables the agricultural industry to create an open-sharing platform that ensures safety for both farming communities and agribusinesses.

## 7.3 Data Security in Agriculture

### 7.3.1 The Importance of Data in Agriculture

The modern agricultural industry has elevated data to become an indispensable element within its operations. Specialized IoT sensors, drones, and multiple remote sensing systems expose information about soils together with agricultural conditions while aggregating climatic and farm operational data. The collected data serves precision agriculture management needs because it leads to enhanced resource efficiency and yield results for farms. Data breaches require security to protect agricultural data, which covers diverse geographical areas among three primary groups: farmers, suppliers, and regulators.

### 7.3.2 Challenges to Data Security

Data security becomes difficult to protect through dispersed agricultural systems. Data protection faces major difficulties since information moves through supply network connections alongside bank institutions, trade hubs, and government agencies. Agricultural system data across worldwide rural areas must move between multiple organizations to face possible security dangers from unauthorized access. Various cross-linking factors within these data-sharing cases prevent the establishment of uniform data-sharing methods while ensuring data security.

The main issue stems from the weak development of rural area infrastructure. Across different regions, there is insufficient availability of secure facilities for data exchange and storage. The lack of secure agricultural data occurs due to insufficient internet connectivity and outdated equipment, while agriculture business operators have limited cybersecurity resources. The implementation of safe data management systems by farmers and agribusinesses depends on adequate infrastructure to reduce the chances of information exposure from improper methods.

Modern digital security threats have made industrial agricultural data security much more difficult to protect. The entire range of farming digital assets encounters hacker attacks that employ digital theft techniques, together with system penetration actions and internet scams for illicit command access. Attacks against agricultural systems become successful due to their lack of necessary security protocols, thus allowing database alterations and personal information exposure through the operational system. Agricultural security procedures lack industry standards because this sector has failed to establish unified security guidelines. The agricultural sector operates without mandatory regulatory standards, thus creating multiple unsecured locations across all operational points. Every industrial sector requires new infrastructure networks that integrate security protocols and cybersecurity systems to develop successful solutions.

### 7.3.3 Applications in Agricultural Data Security

Digital technology adoption by farmers has developed data security into an essential problem. The vital need for secure data management emerges because of combined data production from IoT sensors, automated machinery, and cloud-based platforms [4]. Blockchain technology integration provides agricultural stakeholders with improved data security because the approach establishes visible systems and protected data access while reducing threats against exposure. Blockchain applications operate through technology to establish enhanced protective systems for agricultural data storage. The tracking process of agricultural commodities across supply chains demonstrates a central blockchain application that connects transparent data management techniques to market entry anti-deception mechanisms. Users and organizations can use this system to verify agricultural products alongside their origin, therefore reducing the possibilities of fraudulent products and contaminants.

Secure data transfer from IoT devices gets support from blockchain technology in applications for smart farming. Blockchain security features safeguard all smart irrigation systems and farm sensor output data from unauthorized modification of real-time information. Accurate, trustworthy data, which this practice ensures, allows decision-making processes to boost operational efficiency and farm administration dependability. Users benefit from blockchain technology by receiving superior data-sharing tools that deliver safe information-sharing with multiple stakeholders. A secure system enables farmers, when combined with agribusiness entities and regulatory entities, to share accurate agricultural data through encrypted communication networks. Blockchain security integration delivers elevated privacy protection for stakeholders in agricultural organizations while maintaining their operational cooperation. The agricultural sector offers data security options to stakeholders to build mutual trust through blockchain implementation to provide dependable digital farming systems.

## 7.4 Privacy-Preserving Data Sharing Framework

### 7.4.1 Federated Learning Models

The impact of climate change on commercial grain agriculture becomes severe because of three critical factors that undermine production yields, agricultural sustainability, and quality. The implementation of CSA continues to take hold as a solution to address the discussed problems. The CSA framework enables the transportation of sustainable greenhouse gas control, climate-resilient agricultural production, and effective sustainable productivity as its primary elements. Federated learning (FL) represents a sought-after method for CSA since it enables training teams to work with dispersed data while preserving data security and privacy. Federated learning provides teams with model training capabilities because it executes distributed source operations without giving away data visibility. Individual clients (farmers) train models with their local information at their specific locations, but a centralized server handles the training process through FL. The central server handles model updates through parameter and gradient transfers to protect data privacy.

### 7.4.2 Key Benefits

The advanced machine learning system, FL, enables the distributed model development process without moving original data to a central server. It exists as a decentralized training solution that provides essential services for agricultural applications because farmers need to maintain large sets of sensitive information regarding soil conditions, weather patterns, and crop yield statistics.

- Federated learning enables producers to store their agricultural data locally, which decreases privacy risks to their operations. It enables model training procedures on farmers' devices or local servers, so any sensitive information remains within its original location. The local data storage of FL protects confidentiality and safeguards farmer data, which enables hesitant producers to trust sharing their information externally.
- The FL platform enables better training of resilient and precise models. Federated learning leverages information collected from multiple farms situated worldwide to establish improved and specific predictive models that benefit farmers [5]. Understanding crop yield projections and identifying plant diseases through these models leads to optimal resource management, enhancing farming operations through better decision-making.
- Federated learning operates effectively for businesses that maintain commercial grain farms and large operations. The system demonstrates high scalability because it processes many users together with different types of data sources. Future Learning enables extensive farming businesses to process massive agricultural datasets through their efficient FL security protocols across extensive farming operations.
- The implementation of FL shortens the traffic of server-user data transmission. Federated learning technology does not require the transfer of complete datasets because it

distributes just model update components between a central server and individual nodes. It presents a practical solution because it reduces data transmission expenses and raises system operational speed for areas with restricted internet access.

Federated learning implemented in agricultural data processing enables farmers to gain protected privacy in their information and precise model analytics while cutting operational expenses.

### 7.4.3 Federated Learning Models for Climate-Smart Agriculture

Modern farmers use FL models to implement CSA systems, which brings transformative power to climate-smart agriculture since it enables data-driven decisions through the protected privacy of sensitive data. Remote farmers can access superior analytical capabilities through this system because the platform protects sensitive farm data while bringing advantages to its users. Therefore, this constitutes a perfect modern agricultural solution. Federated learning models serve multiple areas in CSA to enhance operational efficiency and environmental sustainability. Federated learning achieves essential solutions by applying analysis to crop yield prediction. The yielded yield prediction models from FL achieve accuracy by acquiring data from dispersed farm locations covering different geographic territories. The acquired information helps farmers optimize resource distribution while assisting them with choosing optimal harvest periods to reach maximum production rates and prevent crop damage.

Federated learning shows its significance to disease detection through a second vital application. Farmers provide agricultural crop images to FL for disease detection to avoid future problems, so they can prevent damage early on. The utilization of analytical data through preventive procedures lowers crop failures and decreases pesticide volume, supporting sustainable farm operations. Implementing FL enables early disease control, which creates healthy crops and minimizes environmental damage. Agribusiness operations utilize FL to monitor soil health as a critical agricultural application. Federated learning uses multi-source soil sensor data collection to find hidden changes between nutrient density and total soil fitness on agricultural land. Soil management adjustments through FL help farmers reduce fertilizer use, thus preserving long-term sustainability and fertility of agricultural land.

Federated learning models enhance the prediction capabilities for weather conditions. Farmers benefit from FL models, which combine many climate inputs to generate precise weather predictions because these models allow better preparation for severe weather events. Through this capability, farmers control irrigation times, as well as safeguard their crops while making farming decisions optimal for planting. The application of FL within key agricultural zones improves climate-smart farming systems and enhances long-lasting benefits for farmers while maintaining environmental sustainability.

### 7.4.4 Differential Privacy Mechanisms

Differential privacy (DP) is an appropriate framework for data privacy, which will be used in calculations; it ensures that an analysis of an individual's data does not lead to the identification of any person or the leakage of personal data. Because CSA involves data aggregation, DP can be a significant way to keep the farmers' data invisible to others, thus protecting data privacy, while also allowing for the extraction of valuable information from such data. Information about farming, such as crop productivity, soil composition, and pesticide use, is private and sensitive [6]. As a result, DP would both safeguard this data and enable group analysis. By fostering trust between farmers and analysts, data security *via* DP would allow for legitimate and buoyant data sharing and cooperative problem-solving. DP may be able to address issues with data privacy. It can support firms in following privacy regulations such as GDPR and CCPA, keeping them within the law and exempt from prosecution.

Protecting climate-smart agriculture data requires implementing multiple mechanisms that follow DP standards. Query results contain noise additions under the Laplace and Gaussian mechanisms to obscure specific data points but maintain general data usefulness. A random noise implementation to utility-based response selection under the disturbance method creates challenges for specific data extraction. The clipping method helps protect against disproportionate influence by extending beyond the parameters of undesirable extreme values present in datasets. It minimizes their impact on the results. The implemented mechanisms provide secure data analysis for agriculture through safeguarding farmer-sensitive information. The latest information technology with DP capability enables proper, data-driven strategy development for CSA maintenance of personal data security. The technique enables better choices regarding resource distribution, crop management, and sustainability steps, which result in increased agricultural outputs alongside protecting confidential data and security requirements for all participants in the process.

### 7.4.5 Challenges of Differential Privacy for Climate-Smart Agriculture

Differential privacy serves as a usable approach to protect personal information in CSA systems, but it faces multiple obstacles before becoming widely accepted. The fundamental challenge stands between ensuring privacy and maximizing utility [7]. Greater privacy settings tend to reduce accuracy, thus diminishing the value of acquired insights. Processing power consumption rises extensively when implementing DP because of working with substantial agricultural datasets. Its ability to use restricted processing resources in challenging areas is negatively affected. Implementing DP in the demanding and labor-intensive data analysis field, which includes machine learning models, proves challenging because it needs to balance privacy protections with useful information delivery. Understandable knowledge and approval of DP remain essential for users to embrace. The stakeholders, both farmers and others, must understand both their regular safety practices and all the constraints and effects of DP results sharing.

The main difficulty of implementing DP stems from its utility-privacy trade-off. Data privacy specifications create major implementation difficulties because they trigger decreases in data utility, reducing the accuracy of secure information protection. A successful implementation demands a dependable approach to managing privacy requirements against useful data benefits. An excessive number of large data collections makes DP model implementation tedious by demanding considerable processing power and infrastructure. Major resources, along with high implementation costs, become necessary for operating these methods. The DP methods, including local privacy noise, create barriers for complex data analysis due to their use in machine learning and advanced analytical applications, which make practical application challenging. User responses act as an important determinant of whether organizations adopt DP systems. Users need to understand DP limitations because this protocol does not create a completely protected environment against data breaches, yet they must also accept that privacy methods could reduce data-driven insight accuracy and reliability.

## 7.5 Securing Agricultural Data Systems

In the context of CSA, blockchain technology provides a strong solution for maintaining data integrity. Because blockchain is a decentralized, unchangeable ledger, it records data in a way that makes it visual, verifiable, and free of falsifications [8]. This increases product shelf life, which in turn leads to tracking the entire cycle from seeds to the table and promoting openness by making the data publicly verifiable. Furthermore, the blockchain can serve as a platform for sharing information among interested parties, providing regulated access to potentially hazardous data. The smart contract, an automated agreement with terms directly coded, can speed up the lengthy procedure and boost contract sustainability at the same time.

The main topic of discussion here is the fundamental usage of blockchain in tracking crop conditions, soil fertility, pesticide use, and weather data analysis [9]. After being stored on a blockchain, this data becomes unchangeable and can be examined to reveal trends, spot patterns, and ultimately help you make the best choice possible. More significantly, blockchain is utilized to verify the legitimacy of sustainable and organic certifications, allowing customers to verify manufacturers' claims. Blockchain has potential, but several important problems must be resolved, like scalability, energy usage, complexity, and regulatory concerns. By finding the most cutting-edge solutions in fields like layer-2 scaling, energy-efficient consensus processes, simpler interfaces, collaboration, and standardization, these problems can be handled [10]. This can be resolved by addressing these issues, and in fact, a blockchain can transform agriculture for climate control purposes by making it more efficient, transparent, and ecologically beneficial.

## 7.6 Blockchain Can Enhance Climate-Smart Agriculture

Climate-smart agriculture can greatly benefit from increased data transparency and authenticity thanks to blockchain technology. By using blockchain technology to track the food supply chain from farm to table, food safety can be ensured throughout its entire trip, and food quality can be checked at every stage, including origin, processing, transit, and storage.

Transparency is essential for getting goods from the main producers to the customers that need them [11]. The product's sustainability and ethical standards are influenced by the manufacturing process from start to finish, which helps identify and resolve potential supply chain issues. The authenticity and origin of agricultural operations can be verified with blockchain technology. In essence, this ensures that clients receive what they want by blocking fraudulent ones.

Blockchain technology's unchangeability guarantees that the transparent and archived ledger is unchangeable. As a result, it becomes challenging to guarantee the integrity and dependability of the data pertaining to climate-friendly activities, particularly the use of pesticides, water, and carbon footprint, as data is changed or falsified. Smart contracts can ensure that tasks are completed accurately by verifying and executing sustainable practices [12]. For instance, they can ensure that farmers adhere to environmental regulations, such as minimizing pesticide use and cultivating crops with sustainable irrigation methods. Blockchain technology may enable farmers who employ environmentally friendly practices to receive bonuses and incentives. As a result, by introducing the smart farming concept for more environmentally friendly farming, the system encourages a change in agricultural practices.

Another advantage of blockchain is its ability to determine the carbon impact of food items across the supply chain [13]. The precise method that blockchain technology works has the potential to significantly and favorably change sales tracking and monitoring for both customers and food production businesses. The key technologies and their role in addressing data challenges in climate-smart agriculture are given in Table 7.1.

**Table 7.1** Key technologies.

| **Technology** | **Role in addressing challenges** |
|---|---|
| Federated Learning | Enables collaborative model training without sharing raw data, thereby enhancing privacy and security. |
| Differential Privacy | Adds noise to data to protect individual privacy while preserving data utility. |
| Blockchain | This ensures that the agricultural data recorded and shared is safe and transparent, with integrity and traceability. |

## 7.7 Discussions

### 7.7.1 Farmer-Centric Data Ownership Policies

To guarantee both control and benefits from using their data, farmers need to serve as central participants in data ownership legislation. The collection of agricultural data requires prioritizing privacy, alongside security measures, fair payment for farmers, and clear visibility for policy execution. When farmers take possession of their data through encouragement, they activate innovation, which benefits their industry to build a sustainable food system that promotes equity. The generated farm data must belong to producers who should determine their use practices and decide between sharing and monetary value creation [14]. Farmer data privacy, along with security, needs to be strong because it blocks unapproved users from accessing vital information.

Farmers need both operation cost reduction and visible data-sharing program benefits as compensation for their data value. Farmers must access agricultural data that operates between multiple platforms so they can exchange information without interruptions [15]. A governance framework with transparent rules should monitor the ethical handling and integration of farmer-sourced information. The agricultural sector will develop into a sustainable and equal system when it adopts data policies that prioritize farmers as owners. Achieving data control for farmers leads to agricultural innovation while generating new income streams and digital agricultural developments benefit their needs effectively.

### 7.7.2 International Standards for Agricultural Data Security

Internationally, protecting agricultural data is crucial. It is essential for maintaining information confidentiality and ensuring seamless international data sharing. The following are the fundamental international guidelines for agricultural data security:

- Information Security Management System, ISO/IEC 27001: One of the main objectives that addresses every aspect of the information security management system is this standard. It protects against a range of security mechanisms, including incident response, encryption, and ATC Preventive.
- The General Data Protection Regulation: Aiming mainly to monitor personal data, GDPR has some significance to agriculture that cannot be completely disregarded, particularly

if it involves the personal data of farmers or consumers. It pays more attention to data rights, privacy, and acceptance.

- The National Institute of Standards and Technology Cybersecurity Framework: This framework was created by the National Institute of Standards and Technology. A risk-based approach to cybersecurity risk management lies at the heart of the framework [16]. The same concept can be applied to the agricultural industry to identify, evaluate, and reduce cybersecurity risks.
- Standards to the Industry: Industry-specific standards, such as those created by the International Electrotechnical Commission and the International Organization for Standardization, address specific security issues in agriculture, such as data exchange, traceability, and IoT security.

These guidelines promote cooperation between the parties, trust-building, and consistency in data security, privacy, and security practices.

## 7.8 Case Studies: Overcoming Challenges of Data Privacy, Security, and Scalability for Commercial Grain Farming

### 7.8.1 Case Study: Remote Sensing and Geographic Information Systems-Based Crop Monitoring

Technologies like remote sensing and Geographic Information Systems (GIS) have completely changed crop monitoring in India, especially when it comes to commercial grain production [17]. In order to improve agricultural methods and make them more environmentally friendly and climate resilient, this enables real-time evaluation of how the crop is growing and how the conditions are supporting or affecting the crop.

#### *7.8.1.1 Initiation by the Indian Space Research Organization*

India's agriculture and farming industries form a key part of the economy and, as a result, employ almost 60% of the Indian population. Their reliance on monsoon rains and other unpredictable weather changes makes technological advancements in food security and other sustainable farming practices very important. In that regard, agricultural ministries of states, along with the Indian Space Research Organization (ISRO), have created sophisticated crop monitoring systems that utilize GIS and other remote sensing technology. They have shifted expectations around agriculture for policymakers, researchers, and farmers during difficult weather problems. Remote sensing and GIS technology cause a defect in monitoring agriculture to positively enable the collection of significant amounts of data [18]. Remote sensing is the act of collecting data from satellites, while GIS is spatial analysis software. Both technologies report to farmers immediately if their plants need to be monitored, the soil is healthy, the water available to them is adequate, or if the weather is favorable.

The ISRO has initiated a number of programs to utilize satellite technology for agricultural growth. Two of the most well-known programs are Forecasting Agricultural output using Space, Agro-meteorology, and Land-based observations (FASAL) and the National Agricultural Drought Assessment and Monitoring System (NADAMS).

- **Forecasting Agricultural output using Space, Agro-meteorology, and Land-based observations:** Initiated to offer pre-harvest estimates of crop production, FASAL integrates satellite data with meteorological data and field observations to forecast crop yields accurately. It utilizes satellite sensors to track crop growth phases, estimate vegetation indices, and measure factors such as biomass and canopy cover. The crop forecasts by FASAL assist the government in food grain procurement planning, buffer stock management, and timely action against food scarcity concerns.
- **National Agricultural Drought Assessment and Monitoring System:** Formulated to track and evaluate agricultural drought, NADAMS applies satellite data to monitor vegetation stress, soil moisture, and rainfall anomalies. The system offers district-level warnings for drought, allowing state governments to take relief actions, distribute resources, and assist impacted farmers. Reducing the effects of drought, NADAMS has played a significant through early warning systems and directing drought-proofing strategies.

#### *7.8.1.2 Characteristics and Advantages*

Computer-based remote sensing and GIS technology radically enhanced crop monitoring through high-definition imagery of agricultural land.

The GIS-based image analysis allows Cartosat and Resourcesat satellites to capture precise imagery for assessing vegetation growth and field maturity, along with drought and insect-related threats. These modern technologies provide an exact assessment of extensive agricultural areas by gleaning important data that serves farmers and policy masters. The protection of data stands among the most essential components in agricultural information management systems. The sharing of sensitive data operates through multiple encryption protocols, which together limit access to approved users only. The secure servers keep information safe because only authorized users who pass authentication tests can access it, protecting farmers' information from breaches and unauthorized use.

By integrating remote sensing with GIS technology, the system achieves high scalability, enabling proper control of expansive land bases exceeding millions of hectares. Cloud computing utilizes distributed data-processing algorithms to enable the system to quickly process tens of thousands of sensors across multiple sites without declining performance outputs. The system design allows monitoring operations to receive immediate yet accurate information about crop situations. Agricultural decisions through GIS lead to higher operational resource efficiency, which produces more accurate predictions during decision frameworks. Farmers who use spatial analysis tools with stakeholders enhance both land resource management and water program effectiveness, thus improving resource utilization. The advancement in time of remotely acquired yield statistics enables farmers and government agencies to improve their storage and distribution plans, securing food availability and decreasing losses during the post-harvest period.

The systematic combination of mapping systems enables quick identification of water-stressed areas to guide targeted action by improving irrigation systems across these regions. Detecting abnormal vegetation patterns is possible through remote sensing technology because it functions at scales beyond human perception. The immediate detection of problems enables the quick activation of control measures that minimize losses to crops. Market analysis receives its basis from detailed agricultural information, which supports stable food costs and reduces price inefficiencies. Real-time insights applied to supply chains help minimize disruptions, which lead to fair market standards between farmers and consumers. The combination of remote sensing and GIS technologies has developed into essential agricultural tools that help improve productivity and sustainability and strengthen economic resilience.

#### 7.8.1.3 Challenges and Solutions

In India, commercial grain farming has been transformed using remote sensing and GIS for crop monitoring. This has reduced the hazards related to climate variability, enhanced resource usage, and provided farmers with real-time information [17]. By offering trustworthy data for disaster management and food security planning, it has also reinforced governmental policy. Agri-adas processing depends on maintaining data privacy because farm-related records frequently hold confidential information. The solution for this involves restricting access to detailed datasets and using aggregation methods before third parties receive any insights. Protected individual farm data gets preserved through targeted security precautions, enabling beneficial sector-wide agricultural analysis.

Insufficient internet connectivity and IT infrastructure in distant locations present an important barrier to infrastructure requirements. Numerous rural areas lack essential digital resources required to run data-driven solutions at their highest levels. The establishment of ground stations by ISRO, along with enhanced digital networks through their organization, has enabled better access to agricultural data systems. Technical data remains a challenge to farmers since many individuals face difficulties converting statistical information into real-world farm applications. Agricultural extension services, together with state governments, protect against this problem by developing mobile-based advisory platforms that deliver information in regional languages. These programs establish means for farmers to properly translate data-based understandings into effective agricultural procedures. This case study serves as an example of how cutting-edge technologies can support sustainable farming methods while addressing issues of data privacy, security, and scalability.

### 7.8.2 Case Study: e-Choupal by ITC

#### 7.8.2.1 Overview and Implementation

ITC Limited launched the e-Choupal program to help Indian farmers, especially those engaged in commercial grain farming, by removing inefficiencies from the agricultural supply chain [19]. Launched in 2000, this eliminates middlemen by connecting farmers and markets directly with internet-enabled kiosks that provide real-time market pricing, weather, and best agricultural practices. Data privacy leaves much fear of misuse of personal and farm data. The e-Choupal program dispels that worry by anonymizing the collected data and ensuring that usage is always controlled by the individual farmer. The data security system has put in place robust security, such as secure logins and encrypted communications, protecting sensitive transactional and farm-level data from breaches. The e-Choupal program is engineered to deal with the vast diversity and scale of Indian agriculture. It has managed to scale up to cover more than 4 million farmers in 40,000 villages, thus proving it can work at scale.

Internet-enabled kiosks serve as the farmers' interface and are found in villages. They offer digital services to farmers and are run by *sanchalaks*, who are qualified local representatives. Farmers can obtain specific agronomic advice and monitor the daily grain prices at local *mandis* (markets). To avoid unfair price extraction or any form of intermediation, farmers are permitted to sell their produce to ITC.

#### *7.8.2.2 Technology and Security*

Concurrent data protection occurs in the e-Choupal system due to its centralized database management, which implements controlled access systems. The central computer system removes data from individual farmers to combine it with merged information for result sharing among interested parties. No obstacles prevent new service additions or village-based expansions since the platform functions with modular architectural design principles. Fresh market information delivery helps farmers remove middleman agents, which allows them to earn better profits from their agricultural produce. Farmers achieve economic stability by removing middlemen distributors, which both lowers payment expenses and generates new revenue possibilities that help farmers improve their financial situation while enhancing their quality of life.

Agricultural production improvement relies mainly on better education as its fundamental driver. Farmers can generate outstanding field productivity through proper agricultural techniques combined with weather forecasts to develop sustainable agricultural systems. The farmers implement resource optimization strategies after gaining crop-based knowledge, reducing financial losses while achieving definitive operational outcomes. Digital agricultural platforms gain trust through their established pricing systems and protected data security measures, which farmers heavily depend on. These reliable systems earn farmers' trust, ensuring their continuous use of such platforms. Digital agriculture retention of farmers continues over the long term through established pricing systems and protected transactions, which preserve sector value. At the start of digital implementation, most farmers found it difficult to learn digital capabilities.

The company designed widespread training solutions as a response to these difficulties. The company established partnerships with telecom providers because those regions lacked adequate internet connections.

## 7.9 Conclusion

This chapter marks a pivotal moment for commercial grain farming, which presents major opportunities and significant challenges in the era of digital transformation. Modern farming practices have become more efficient and sustainable by implementing cutting-edge technologies like the IoT, blockchain, AI, and remote sensing. These technological advancements have surfaced critical issues around data privacy and security, as well as scalability, which require proper resolution to enable successful CSA practices. Protecting data privacy enables farmers to control their confidential information while implementing governance policies combined with privacy-preserving technologies like FL and anonymization, establishing trust to enable digital agricultural tool adoption on a broad scale. Agricultural data requires robust protection measures such as encryption and blockchain solutions to maintain operations and livelihoods, as data breaches can pose serious threats.

**AI disclosure statement:** The author confirms that no artificial intelligence (AI)-generated content or AI-assisted writing tools were used in the preparation of this chapter (including summaries and the abstract). All text, analysis, and conclusions are the result of the author's original work.

## References

1. Lucena, P., A Case Study for Grain Quality Assurance Tracking Based on a Blockchain Business Network, *arXiv preprint arXiv,1803.07877*, 2018.
2. Yogeshwari, M. and Prasanth, A., Smart Dust Technology for Monitoring and Control Systems in Smart Agriculture and Crop Surveillance Systems. *Com. Vis. Smart Agric. Crop Man.*, 123, 137, 2025.
3. Gandhi, R., Digital Green: Participatory Video and Mediated Instruction for Agricultural Extension. *Inf. Technol. Int. Dev.*, 5, 15, 2009.
4. Abinaya, M. and Vadivu, G., Role of Big Data Analytics in Intelligent Agriculture. *Comp. Intell. Int. Agric. Things*, 12, 19, 2024.
5. IFFCO Kisan, Enabling Digital Solutions for Farmers, 2022, Retrieved from https://iffcokisan.com.

6. Carolan, Publicising Food: Big Data, Precision Agriculture, and Co-Experimental Techniques of Addition. *Sociol. Rural.*, 57, 135, 2017.
7. Lin, W., Blockchain Technology in Current Agricultural Systems: From Techniques to Applications. *IEEE*, 8, 143920, 2020.
8. Arulkumar, V., Krishna, M., Jayachitra, S., A Future Cyber Security For Secured Smart Home Applications Using The Internet of Things and Blockchain Technology. *Multi. Sec.*, 17, 202, 2025.
9. Kamilaris, A. and Prenafeta-Boldú, F.X., The Rise of Blockchain Technology in Agriculture and Food Supply Chains. *Trends Food Sci. Technol.*, 91, 640, 2018.
10. Van der Burg, S., Ethical Considerations in Data-Driven Agriculture: The Case of Smart Farming. *NJAS Wagening. J. Life Sci.*, 90, 100289, 2020.
11. Lobell, D., Climate-Smart Agriculture for Food Security. *Nat. Clim. Change*, 3, 4, 2013.
12. Khan, M.A. and Salah, IoT Security: Review, Blockchain Solutions, and Open Challenges. *Future Gener. Comp. Sy.*, 82, 395, 2018.
13. Ge, L., Blockchain for Agriculture and Food: Findings from Pilot Studies. *J. Econ. Res.*, 21, 245, 2017.
14. FAO and amp: ITU, E-Agriculture in Action: Blockchain for Agriculture, 2019, Retrieved from https://www.fao.org.
15. Farooq, M.S., A Survey on the Role of IoT in Agriculture for the Implementation of Smart Farming. *IEEE*, 7, 156237, 2019.
16. Gollin, D., Digital Agricultural Services for Smallholder Farmers. *Annu. Rev. Resour. Econ.*, 13, 317, 2021.
17. Ministry of Agriculture and amp; Farmers Welfare, Government of India, AgriStack: Building a Digital Ecosystem for Indian Agriculture, 2021, Retrieved from https://agricoop.nic.in.
18. Indian Space Research Organisation (ISRO), Applications of Remote Sensing in Agriculture: FASAL and NADAMS Programs, 2020, https://isro.gov.in.
19. Ray, S. and Bhattacharya, S., e-Choupal and Its Impact on the Social and Economic Empowerment of Farmers in India. *Int. J. Rural Manag.*, 7, 45, 2011.

8

# Seizing Opportunities in Integration of Reinforcement Learning with the Internet of Things for High-Tech Greenhouse and Vertical Farms

**N. Sathish[1]*, V. Yokesh[2], Prasanth Aruchamy[3] and Pham Chien Thang[4]**

*[1]Department of Electronics and Communication Engineering, Sri Venkateswara College of Engineering, Sriperumbudur, India*
*[2]Department of Electronics and Communication Engineering, Sathyabama Institute of Science and Technology, Chennai, India*
*[3]Department of Computer Science and Engineering, Vel Tech Rangarajan Dr. Sagunthala R&D Institute of Science and Technology, Chennai, India*
*[4]Faculty of Journalism, Communication, and Literature, Thai Nguyen University of Sciences, Thái Nguyên, Vietnam*

***Abstract***
Two prominent models for enhancing vertical farming and smart greenhouses are reinforcement learning (RL) and the Internet of Things (IoT). This section explores how various sources of man-made IoT systems can be combined with RL to automate and optimize processes applied to sustainable agriculture systems. Reinforcement learning is known for its adaptive learning capabilities, which are important in the complex environments of a greenhouse or vertical farm. It can ensure speedy adjustments to optimal growth by its responsive control over several conditions. Additionally, RL can quickly adapt to the highlighted growth conditions and manage resources accordingly. Utilizing IoT networks with high-accuracy sensors for continuous observation to dynamically adapt RL weather prediction algorithms to environmental changes in light, humidity, temperature, and soil quality. This combination of technologies paves the way for precision farming, promoting higher performance and lowering price points, and a platform for greater optimization of the entire value chain. This chapter uses

**Corresponding author*: n.satyish@gmail.com; ORCID: 0009-0000-3507-6337

Anitha Velu, Prasanth Aruchamy, Raghu Ramamoorthy, Rajesh Kumar Dhanaraj and Seifedine Kadry (eds.) Climate Smart Agriculture, (147–172) © 2026 Scrivener Publishing LLC

RL and IoT to address modern agricultural problems like personnel shortages, climate change, and organic farming. It also handles data management, computational demand, and system integration issues while extending those technologies.

***Keywords***: Reinforcement learning, smart agriculture, Internet of Things, greenhouse, vertical farms

## 8.1 Introduction

### 8.1.1 Overview of High-Tech Greenhouses and Vertical Farms

High-tech greenhouses and vertical farms are the poster children of modern agricultural innovation, using a cocktail of recent innovations to expand plant growth, resource use efficiency, and production. These systems are designed to address problems such as climate change, land availability, and the increasing demand for sustainable food production.

- **High-Tech Greenhouses**: Buildings are controlled-environment buildings where the temperature, humidity, light, and levels of $CO_2$ can be controlled by technology. Through the combination of automated systems and IoT-enabled devices, they ensure an optimal environment for plant growth with minimal resource waste.
- **Vertical Farms**: Vertical farms feature multi-story buildings that allow for maximized use of space; they are typically located in urban regions. With advanced hydroponics, aeroponics, and precise fertilizer delivery systems, you get more food with less water and land.

Key features include:

- Precision agriculture techniques for tracking and controlling growth parameters
- Integrating renewable energy to enable sustainable operations
- Predictive maintenance and growth optimization through AI-driven insights

These breakthroughs are key to future food security, unlocking the potential for a sustainable and resource-efficient agricultural ecosystem.

### 8.1.2 Role of the Internet of Things in Modern Agriculture

The IoT enables modern agriculture to conduct real-time monitoring, data-driven decision-making, and improved operational efficiency [1] (Figure 8.1). The IoT devices include sensors, actuators, and connected networks that facilitate the adoption of technology into agricultural practices and provide farmers with valuable data.

- ***Precision Monitoring:*** The IoT sensors are used to measure temperature, humidity, light intensity, and soil moisture. By controlling growing conditions at high levels of precision, this data minimizes waste and increases crop quality.
- ***Automated Operations:*** Ten IoT-enabled technologies automate critical chores like pest management, fertilization, and irrigation. These automated processes ensure that farm operations are performed in a timely manner and reduce human involvement.
- ***Data Collection:*** With IoT devices, huge amounts of information are collected, and this information could be analyzed to predict yields, optimize the use of resources, and gauge fruit health. Unstructured data is transformed into valuable knowledge that assists farmers in making better

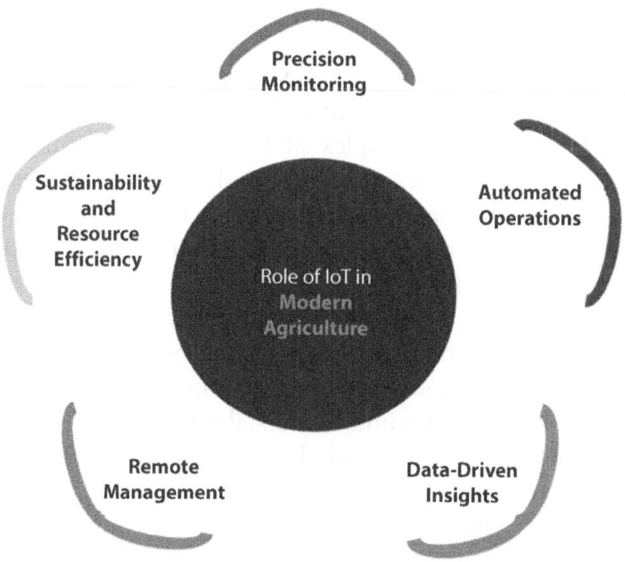


**Figure 8.1** The role of IoT in modern agriculture.

decisions using machine learning (ML) algorithms and big data analytics.

- ***Remote Management:*** IoT connectivity gives farmers the freedom and efficiency to remotely monitor and manage farming activities through web interfaces or mobile applications.
- ***Sustainability and Resource Efficiency:*** The IoT provides a streamlined way to convert energy, water, and fertilizers into production, allowing farmers to achieve their sustainability goals. With IoT, farming operations can produce more while wasting less and staying profitable in the long run.

The incorporation of the IoT in farming also addresses key challenges such as labor shortages, climate variability, and the need for sustainable practices, as well as improving farm production. It is a key part of all of this, unlocking the potential of vertical farms and high-tech greenhouses.

### 8.1.3 Potential of Reinforcement Learning in Smart Farming

Inserting the possibility of learning, adaptation, and development with respect to a shifting environment over time in agricultural (and livestock) systems using one of the power game-changing aspects of RL [2], it can be conceived as one of the exciting future possibilities of agricultural and livestock applications, and has been one of the central themes of the present dissertation research.

In contrast to traditional program schemes, RL develops decision-making and resource allocation through embodied interactions with the environment. Finally, for complex and stochastic agricultural systems (e.g., heterogeneous climate, epidemics, and consumer requests), the RL algorithms themselves can be applied. Indeed, RL agents can be used not only to select the best response in an expected scenario (and thus also efficiency gains due to learning can be realized in such scenarios). Every RL framework aims to partition fertilizer, electricity, and water as fairly as possible for optimization. The RL systems can be designed in such a way that resources are minimized, and green actions are encouraged through a proper balance between proximal and distal rewards.

Due to its ability to continuously learn and adapt to dynamic external meteorological data changes, RL can also be applied to indoor temperature, humidity, and light control in so-called high-tech indoor environments (e.g., greenhouses). Thus, the best growth conditions to carry out the growth process are maintained, together with an increase in energy savings. By analyzing environmental information and plants' health indices, RL models could also be used to predict and take preventive measures for disease outbreaks and plant pest infestations. Prompt action guarantees healthier crops and lessens crop damage.

Reinforcement learning enables self-sufficient systems to undertake actions such as robotic pruning, weeding, and harvesting. These solutions are, to an extent, self-learning, allowing farmers to operate more efficiently and respond to the changing environment over time. The implementation of RL with IoT enables farmers to predict crop output more accurately and enhance their agricultural methods to maximize yield. Although smart farming with RL is still in its infancy, it has huge possibilities. With its capabilities to enhance sustainability, adapt to changing conditions, and optimize processes, it is a powerful tool for the future of farming, protecting the environment.

### 8.1.4 Objectives and Scope of the Chapter

This chapter investigates the usage of RL and IoT technologies in vertical farms and high-tech greenhouses. The following are the goals:

- ***To determine the present difficulties***: RL and IoT can have a big impact on smart farming and other points by assessing the current constraints.
- ***To determine the present challenges***: Current constraints can be pointed out where RL and IoT might have significantly carried out the advancements.
- ***To provide a wholesome framework:*** A comprehensive architecture for optimization of agriculture that can use a combination of RL algorithms integrated with IoT devices.
- ***To showcase real-world use cases:*** Provision of real-world cases in automatic irrigation, climate control, and pest management, demonstrating the efficiency of IoT-based RL systems.
- ***To assess system performance:*** Create performance indicators and standards to gauge how well integrated RL-IoT systems work in smart farming.
- ***To emphasize upcoming opportunities:*** Talk about new developments, obstacles, and chances to expand RL-IoT integration in agriculture.

This chapter covers the theoretical underpinnings, real-world applications, and potential future paths of RL and IoT convergence in agriculture. It serves as a manual for scientists, engineers, and legislators who want to develop smart farming. So, readers will gain information on how RL and IoT can improve the functionalities of the modern advanced greenhouse and vertical farming to make smart agriculture in the future.

## 8.2 Background and Related Works

### 8.2.1 The Internet of Things in Agriculture: Current Trends and Challenges

Adopting IoT in agriculture has introduced transformative capabilities [3] but also presents several challenges.

#### *8.2.1.1 Current Trends in the Internet of Things for Agriculture*

Several revolutionary phenomena have been brought about by the integration of IoT in agriculture. There is also real-time control over environmental factors for plant growth, namely, soil pH, water content, temperature, and light intensity, among others. Other access technologies, such as the long-range wide area networks and Zigbee, have also advanced connectivity mechanisms, making devices perform remarkably over large regions. Also, at present, artificial intelligence (AI) systems analyze data collected by IoT devices while offering information that makes the decision-making process more intelligent. Some farming commitments are now more efficient due to automation driven by robotics and drones. In addition, sustainability is one of the most vital concerns nowadays; various IoT-based solutions implement ecological approaches and reduce the consumption of resources.

#### *8.2.1.2 Challenges Relative to the Implementation of the Internet of Things*

However, many challenges can be found in adopting IoT in agriculture as follows. The first of these difficulties is integrating IoT systems for different farming practices, geography, and land topography. To many farmers, the initial capital investment, such as constructing structures, may be too expensive. Among the data security and privacy issues, data security raises a concern and identifies the need to protect such information.

Moreover, this is due to the various IoT devices lacking harmonized standard operating protocols. Finally, the last one is the ability to keep a device working steadily in unfavorable outdoor conditions, and its reliability. However, before realizing the full potential of IoT in enhancing the agricultural value chain, these challenges must be addressed.

### 8.2.2 Fundamentals of Reinforcement Learning

Reinforcement learning is a type of ML in which an agent must make decisions by interacting with an environment. The problem is to select the optimal behavior from observations to achieve the optimal total lifetime reward. However, RL has one major learning technique that separates it from other learning methods: trial-and-error and feedback techniques. A decision-maker is called an agent; the world or space the agent operates in, providing reward feedback, is called the environment, as applied in RL [4]. Commands are the set of options that the agent can execute in the environment, while the state is a snapshot of the environment at any time. Whereas policies encode state–action pairs to optimize gains, gains demonstrate the consequence of an action. The value function estimates long-term rewards for states or state-action pairings.

Some essential RL methods consist of deep Q-networks, which mesh up Q-learning with deep neural networks that help in handling large state spaces, and Q-learning as a model-free learning approach where the agent is trained on what actions to perform in given states. Actor-Critic methods explore a combination of value function and the policy gradient techniques strive for improved results, whereas the functions that optimize policies are called policy gradient methods. Apart from agriculture, RL has applications in self-driving cars for decision-making and navigation, robots for motion planning and control, and video games in training AI to perform beyond biological human capabilities in games such as GO and chess.

## 8.3 System Architecture for the Intelligence-of-Things-Driven Reinforcement Learning in Agriculture

### 8.3.1 Overview of Integrated Systems

In IoT-driven RL [5], integrated systems integrate algorithms, software, and hardware to make the best decisions in challenging situations.

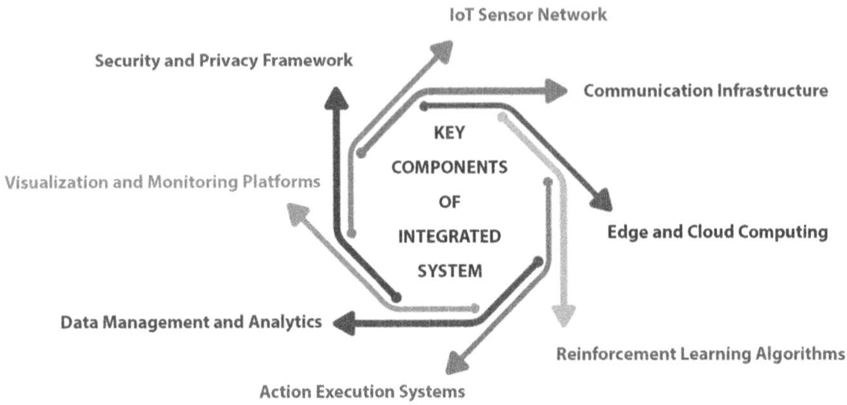


**Figure 8.2** Essential elements of a unified system.

These systems are made to continuously learn from and adjust to changing circumstances while collecting, processing, analyzing, and acting upon data in real-time. Figure 8.2 illustrates the essential elements of a unified system, and Table 8.1 discusses the elements of a unified system.

### 8.3.2 Internet of Things Framework for High-Tech Greenhouses and Vertical Farms

Each of the three main parts of the structural design performs a specific function. The physical structure is the first part of the system. It consists of three parts: microcontrollers, sensors, and actuators. To gather information regarding internal environmental factors, including illumination, temperature, humidity, gas levels, and soil moisture, sensors are required. These sensors transmit the collected information to a microcontroller, which processes data to make decisions. To carry out the necessary operations, the microcontroller communicates with the user remotely over the internet. Actuators are crucial components of greenhouse farming and include things like a nutrient tank, water pump, lighting, and windows. They carry out the user-provided commands.

Monitoring [6, 7], which includes soil, plant, and climatic monitoring, is the second essential element of smart greenhouse farming. Using equipment to maintain the ideal conditions needed for particular crops cultivated in greenhouses is known as climate monitoring. The microcontroller is notified to turn on devices that bring the inside environment back into balance whenever climatic parameters such as temperature, $CO_2$ levels, humidity, or air pressure deviate from user-defined thresholds.

**Table 8.1** Elements of a unified system.

| | |
|---|---|
| **Internet-of-Things Sensor Network** | It is intended to gather a variety of real-time environmental data so that RL algorithms can make defensible choices. The network's interconnected sensors, gateways, and actuators form the basis for putting data-driven, efficient agricultural methods into practice. |
| **Communication Infrastructure** | This forms the foundation of an agricultural RL system powered by the IoT. To facilitate real-time monitoring, decision-making, and action execution, it guarantees smooth data flow between sensors, gateways, computer layers, and actuators. |
| **Edge and Cloud Computing** | These are essential parts of IoT-driven agricultural RL systems that facilitate effective data processing, scalability, and decision-making. Cloud computing offers centralized, scalable resources for advanced analytics and model training, while edge computing offers localized, low-latency processing. These two computing paradigms work in tandem. |
| **Reinforcement Learning Algorithms** | Reinforcement learning algorithms facilitate intelligent decision-making on determining the best course of action to accomplish particular objectives. The agent interacts with environments like crops, soil, and weather in optimizing a cumulative reward, such as crop production, water efficiency, or pest control. |
| **Action Execution System** | It is an integral part of the IoT-enabled RL setup, translating the decisions made by RL algorithms into tangible farming activities. This system features controllers, actuators, and a feedback system for precision and adaptability. |
| **Data Management and Analytics** | The key component for IoT-enabled RL systems in agriculture resides in proper data handling and analysis. To engineer appropriate strategies and improve the system, these steps include collection, processing, storage, and data analysis. |

*(Continued)*

**Table 8.1** Elements of a unified system. (*Continued*)

| | |
|---|---|
| **Visualization and Monitoring Platforms** | This serves as an interface between interested parties, including farmers and agronomists, as well as managers and the IoT-enabled RL system. Such systems facilitate rapid decision-making and improve system visibility by enabling control of agricultural activity, as well as data collection and real-time analysis. |
| **Security and Privacy Framework** | Since the information obtained from sensors and devices integrated with other sources can be sensitive and crucial, it is important in IoT-enabled RL systems for farming. It is imperative to secure communication, system, data, as well as components, to maintain the integrity of the agricultural system and prevent cyber threats such as data leaks and illegal access. |

In soil monitoring, sensors check for salt and moisture, as well as vital nutrients like potassium and phosphorus, to keep plants safe. Crop disease detection and leaf temperature measurement are the main goals of plant monitoring to evaluate plant health and growth. Users can use this information to plan future greenhouse farming cultivation operations and make well-informed judgments regarding plant health, as shown in Figure 8.3.

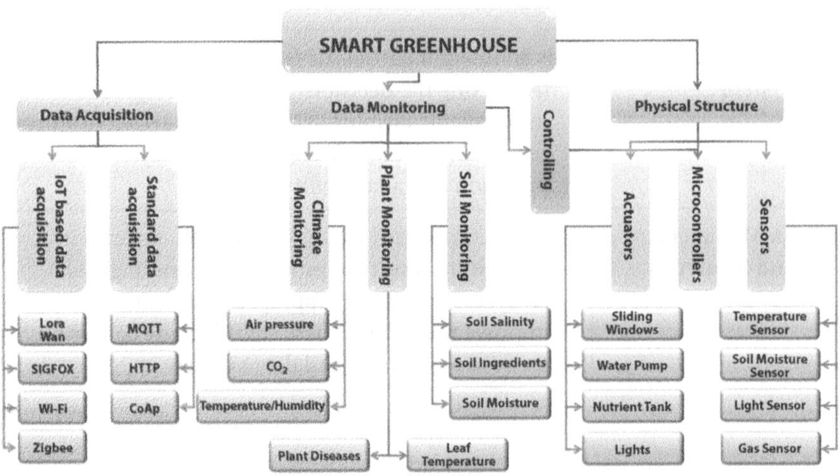


**Figure 8.3** Key components of a smart greenhouse system.

The IoT-based data procurement utilizes the following primary protocols: MQTT, HTTP, and CoAP. These protocols facilitate far-off communication and are instrumental in regulating greenhouse conditions. By optimizing the environment, IoT-based data acquisition helps enhance the efficiency and productivity of high-quality crop cultivation.

### 8.3.3 Reinforcement Learning Framework

In the subset of ML known as reinforcement learning [8], an agent gains decision-making skills by acting in a way that maximizes the total number of rewards in its environment. Reinforcement learning can be used to automate and optimize a number of processes in high-tech greenhouses and vertical farms, including lighting, nutrient delivery, climate control, irrigation, and insect control, as shown in Figure 8.4.

#### *8.3.3.1 Environment*

The operational setup of the greenhouse or vertical farm, including all sensors, actuators, and dynamic factors, is represented by the environment.

- **Temperature and humidity sensors** track the ambient conditions to ensure optimal growth settings.
- **$CO_2$ and $O_2$ sensors** measure gas concentrations critical for plant respiration and photosynthesis.
- **Light intensity and spectrum sensors** monitor and control the lighting conditions to optimize photosynthesis and energy usage.
- **Soil moisture and nutrient sensors** provide real-time data on water levels and nutrient availability to prevent under-saturation or over-saturation.

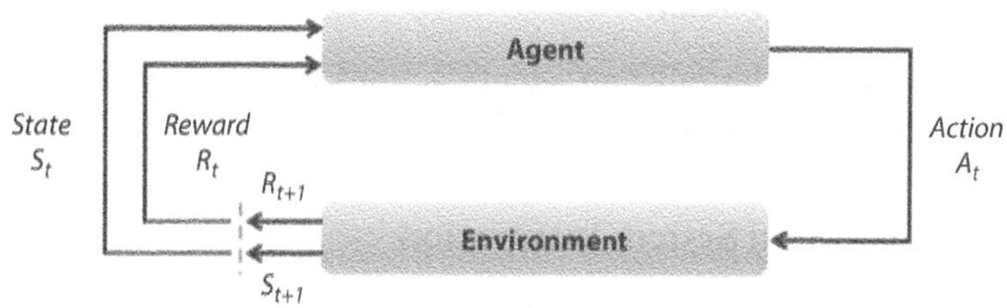


**Figure 8.4** Reinforcement learning process.

- **pH and electrical conductivity sensors** ensure the chemical balance of the soil or nutrient solution is within the desired range.
- **Cameras for visual monitoring** enable detailed observation of plant health and growth, aiding in early detection of pests, diseases, or other abnormalities.

#### *8.3.3.2 Agent*

To accomplish desired goals, the RL agent makes decisions and interacts with the environment by taking actions based on the current state.

#### *8.3.3.3 State Representation*

The state should thoroughly document the current situation of the farm or greenhouse. Usually, it consists of:

- **Environmental Parameters:** Current temperature, humidity, $CO_2$ levels, light intensity, etc.
- **Crop Status:** Growth stage, health indicators, nutrient levels
- **Resource Status:** Water levels, nutrient reservoirs, energy consumption
- **Historical Data:** Past actions and their outcomes for temporal awareness.

#### *8.3.3.4 Action Space*

It defines the set of possible actions the agent can take. Actions might include: adjusting HVAC settings (e.g., increasing/decreasing temperature), modulating irrigation rates or schedules, changing lighting intensity or spectrum, altering nutrient delivery rates, or deploying pest control measures.

#### *8.3.3.5 Reward Function*

By measuring how desirable an agent's actions are, the reward function directs learning toward the best possible policies.

### 8.3.4 End-to-End System Integration

To maximize operations, a high-tech greenhouse or vertical farm must seamlessly integrate multiple components. To facilitate automation, decision-making, and real-time monitoring, the system combines sensors, controllers, actuators, communication networks, and data analytics platforms [9].

Temperature, humidity, $CO_2$ levels, soil moisture, nutrient levels, pH, and visual inputs are just a few of the vital environmental and crop-related data that sensors harvest. Actuators use processed data to carry out control actions in systems including heating, ventilation, and air conditioning, irrigation pumps, LED lighting, nutrient delivery, and shade mechanisms. While a strong data communication network—using wired protocols like Ethernet, Modbus, or CAN bus and wireless protocols like Wi-Fi, Zigbee, LoRa, or Bluetooth ensures smooth data exchange, a centralized or distributed controller unit, such as programmable logic controllers or edge devices, processes sensor inputs and decides what needs to be done.

For real-time analysis and long-term trend monitoring, data is kept on local or cloud servers. Machine learning models and reinforcement learning techniques are used to optimize decisions. Operators may obtain actionable information, monitor performance, and override automatic actions on dashboards or mobile applications. Continuous data collection and aggregation, real-time processing for decision-making, accurate control and actuation, anomaly detection and alarms, and extensive data storage for insights and predictive maintenance are all components of the integration workflow. The IoT facilitates connectivity, edge computing allows low-latency processing, cloud computing manages data, and ML ensures continuous optimization, all of which are essential technologies that ensure the system's correctness, scalability, and efficiency.

## 8.4 Key Applications and Use Cases

### 8.4.1 Climate Control and Energy Optimizations

Efficient temperature management and energy optimization are essential for maintaining optimal growing conditions and reducing operational costs in high-tech greenhouses and vertical farms [10].

#### *8.4.1.1 Climate Control Strategies*

The objective of temperature control strategies in vertical farms and advanced greenhouses is to sustain optimal growing conditions through the use of cutting-edge technologies and real-time monitoring [11]. Advanced heating, ventilation, and air conditioning systems that adjust heating or cooling based on sensor feedback are employed to control temperature. To avoid problems like mold, pests, or plant stress, humidity levels are closely monitored and controlled using dehumidifiers, misters, or ventilation systems. Photosynthesis is improved by controlled $CO_2$ enrichment, and real-time sensors ensure $CO_2$ levels are effectively maintained without waste. Light management maximizes photosynthesis in accordance with plant development cycles by using LED grow lights with programmable spectrum and intensity to mimic natural conditions. To promote healthy plant development, air circulation systems also ensure regular airflow and minimize hotspots by distributing heat and $CO_2$ uniformly.

#### *8.4.1.2 Energy Optimization Techniques*

Smart energy management systems in vertical farms and high-tech greenhouses employ IoT-enabled energy monitoring to pinpoint energy-intensive processes and make necessary adjustments [10]. By delivering light wavelengths suited to various phases of plant growth, energy-efficient LEDs lower electricity use. To improve environmental sustainability, renewable energy sources, including solar panels, wind turbines, and geothermal systems, offer sustainable power for climate control and illumination. Systems for thermal energy storage reduce reliance on external heating by storing heat produced during the day for use at night. Demand response programs also reduce costs and promote grid stability by automating changes in energy usage during peak hours.

### 8.4.2 Automated Irrigation and Nutrient Management

It is crucial for enhancing water and resource efficiency in advanced greenhouses and vertical farms. These systems distribute precise water and fertilizer according to plant requirements by employing sensor networks, advanced algorithms, and IoT technologies. The essential components of automated irrigation and nutrient management are listed in Table 8.2.

### 8.4.3 Pest and Disease Management

To protect crops while reducing chemical use and environmental impact, pest and disease management in high-tech greenhouses and vertical farms depends on integrated methods and cutting-edge technologies. The essential components of pest and disease management are listed in Table 8.3.

**Table 8.2** Essential components of automated irrigation and nutrient management.

| Components | Workflow | Optimization | Benefits |
|---|---|---|---|
| Soil Moisture, Nutrient Sensors, Drip Irrigation Hydroponic Systems, Centralized Controllers, Storage Tanks, and Mixing Units | Data Collection<br>Real-Time Processing<br>Targeted Application<br>Feedback Loop | IoT Integration<br>ML Algorithms<br>Sustainability Features | Promoting healthy growth<br>Reduces water and fertilizer wastage, lowering operational costs<br>Adapts easily to both small-scale and large-scale farming operations<br>Minimizes manual intervention<br>Supports eco-friendly farming practices |

**Table 8.3** Essential components of pest and disease management.

| Components | Workflow | Optimization | Benefits |
|---|---|---|---|
| Cameras and Visual Sensors, Environmental Sensors, Bio-Sensors, Automated Traps and Barriers, Biological Control Agents, Automated Dispensers | Continuous Surveillance, Data Processing Alerts and Recommendations | Precision Spraying, Climate Management, Integrated Pest Management (IPM), and Predictive Models | Early Detection and Prevention, Reduced Chemical Dependency, Cost Savings, Improved Crop Health and Yield Environmental Sustainability |

**Table 8.4** Essential components of crop yield prediction.

| Components | Workflow | Techniques | Benefits |
|---|---|---|---|
| Data Collection Sensors, Data Analytics Platforms, Automation Tools | Data Acquisition, Real-Time Analysis Intervention and Optimization Feedback Loop | Precision Agriculture Practices, Climate and Light Optimization, Nutrient Management, Disease and Pest Prevention, Genetic and Breeding Insights | Higher Yields, Resource Efficiency, Improved Planning, Sustainability |

### 8.4.4 Crop Yield Prediction and Enhancement

Utilizing data-driven insights, sophisticated sensors, and automation technologies, crop production prediction and enhancement in high-tech greenhouses and vertical farms maximizes productivity while guaranteeing resource efficiency. The essential components of crop yield prediction are listed in Table 8.4.

### 8.4.5 Resource Management in Vertical Farming

The effective use of water, energy, nutrients, and space is the main goal of resource management in vertical farming [12] to increase output and reduce environmental impact. High yields are maintained while ensuring the best possible use of resources through applying sustainable methods and advanced technologies.

#### *8.4.5.1 Resources in Vertical Farming*

- **Water:** Recycling systems in vertical farming recapture and reuse water from irrigation processes, significantly reducing wastage and promoting sustainability. When compared to conventional farming practices, advanced hydroponic and aeroponic systems can save up to 90% on water usage by precisely delivering water to plant roots.

- Real-time monitoring with sensors ensures optimal water management by tracking water quality, flow rates, and levels, preventing overuse or shortages and maintaining a consistent supply for efficient plant growth.
- **Energy:** LED lighting in vertical farming, utilizing tailored light wavelengths for photosynthesis, provides energy-efficient illumination and significantly reduces electricity use. Operations utilize renewable energy sources such as wind turbines and solar panels, enhancing sustainability and diminishing reliance on non-renewable energy. By identifying high-consumption activities and adjusting usage during peak periods, IoT-enabled smart energy systems enhance energy distribution and ensure efficient and cost-effective resource management.
- **Nutrients:** Nutrient management in vertical farming emphasizes the accurate provision of tailored nutrient solutions to meet crop requirements, ensuring efficient absorption and minimizing waste. Recapturing, purifying, and reusing nutrient-dense water reduces input expenses and environmental repercussions while promoting sustainability.
- **Space:** Vertical farming optimizes vertical space by stacking to enhance crop output per unit area. Flexible modular systems provide efficient reconfiguration to accommodate diverse crop kinds and growth stages. The AI-based optimization algorithms enhance productivity by determining the ideal planting density and arrangement to maximize yields.

#### *8.4.5.2 Resource Management Strategies*

- **Integrated Monitoring System:** Monitors continuously the resource usage and offers insights from the IoT platform and sensors.
- **Automation and Control:** Usage of real-time data by automated systems to control Irrigation, lighting, and nutrient supply.
- **Predictive Analysis:** Machine learning models anticipate resource requirements and avoid the overusage or shortages of dynamically modified activities.
- **Waste Management:** For the support of a circular economy, plant-based organic waste management has been established.

**Table 8.5** Challenges faced in integrating RL with IoT vs solutions.

| Challenges | Solutions |
|---|---|
| For small farmers, the price of cutting-edge technologies like automation equipment, IoT systems, and renewable energy sources may be unaffordable. | Both public and commercial institutions can provide grants, low-interest loans, and subsidies. Entry obstacles may be decreased by collaborative agricultural infrastructure and lease arrangements. |
| It is technically difficult to integrate sensors, actuators, and control systems into a single, cohesive framework. | Use plug-and-play scalable systems and established communication protocols (like MQTT and Modbus). Working together with knowledgeable technology suppliers helps guarantee effective system configuration. |
| Operational costs are raised by high energy usage for automation, climate control, and lighting. | To maximize power use and cut expenses, integrate smart energy systems, renewable energy sources like solar panels, and energy-efficient LEDs. |
| It can be challenging to manage colossal volumes of real-time data from IoT devices while maintaining data security. | Utilize cloud-based solutions equipped with comprehensive cybersecurity measures, including safe access protocols and encryption. Furthermore, edge computing may alleviate the burden on data storage and transmission. |
| It can be challenging to maintain sophisticated systems and ensure operators can use them. | Regularly train employees, put predictive maintenance systems in place, and provide remote assistance using AI-based diagnostics and digital tools. |
| During shortages, reliance on constant sources of energy, water, and nutrients is risky. | Reduce reliance on outside resources by implementing renewable energy sources, water, and nutrient recycling systems. |
| It can be difficult to win over customers and scale operations to satisfy market demand without sacrificing quality. | Prioritize open procedures, certifications, and the use of technology to ensure constant manufacturing quality. Phased expansions and modular designs guarantee scalability without sacrificing effectiveness. |

## 8.5 Implementation Challenges and Solutions

To ensure operational efficiency, scalability, and sustainability, the implementation of advanced greenhouses and vertical farms must address several challenges that require innovative solutions (Table 8.5).

Addressing these challenges with targeted solutions would enhance the accessibility, efficacy, and importance of vertical farms and advanced greenhouses, hence promoting sustainable agricultural practices and food security.

## 8.6 Evaluation Metrics and Performance Analysis

Evaluating vertical farms and high-tech greenhouses entails determining overall productivity, resource optimization, and system efficiency. A thorough performance analysis requires a mix of IoT-based measurements, RL-based optimization benchmarks, and comparative analysis.

### 8.6.1 Performance Metrics for the Internet of Things

The performance evaluation of IoT systems in high-tech greenhouses and vertical farms can be done based on several significant benchmarks. Enabling consistent information for decision-making, data accuracy measures the degree of truth or reliability of the measures recorded by the sensors. In real-time updates, latency refers to the time between when data is being taken and the time the system responds [13]. To promote energy efficiency, energy consumption quantifies how much electricity IoT devices consume. Scalability measures the ability of a system to integrate and function within larger numbers of interconnected devices and data feeds. A coverage that checks data communication success rates and system availability guarantees network reliability, which denotes constant and stable network functionality.

### 8.6.2 Reinforcement Learning-Based Optimization Benchmarks

An evaluation of the RL optimization algorithm is made based on certain criteria. The time taken for the RL model to learn and make sound decisions is calculated by the convergence speed. Reward optimization compares the total reward earned in a process and looks beyond the system efficiency and resource utilization [14]. Adaptability measures the algorithm's behavior in response to varied conditions outside the algorithm or any constraints in the system. The temporal cost and the computational effort needed for training and deploying RL models are analyzed by computational efficiency.

### 8.6.3 Comparative Analysis of the Internet of Things-Reinforcement Learning Systems

Comparative examination determines to what extent RL-based optimization strategies work together with IoT platforms in several configurations. They identified and quantified water, energy, and nutrients savings to evaluate resource utilization efficiencies. Yield improvement examines this area as relates to the quality and quantity of the produce [15]. The increase in the use of IoT-RL integrations over traditional systems in terms of operational costs is used to compare feasibility. The system response measures the system's dependability and flexibility in real-time. The ability to spread over a larger surface, increase their scope, or change to meet different circumstances is referred to as scalability and flexibility.

### 8.6.4 Insights from Experimental Results

The IoT experiments are insightful, and so is integrating RL into the system experiments. Resource optimization showcases up to 30% savings in energy, water, and nutrients. The increase in yield establishes consistent improvements in production due to the correctly estimated alteration and dynamic control. Consequently, it is shown that RL algorithms are effective in dynamically optimizing environmental conditions, as evidenced by better decision-making. As supported by the economic feasibility assessment, cost-saving is accomplished through atomization, as well as efficient utilization of available resources [16]. The general scalability of these systems is proven by their capacity to operate in different contexts and scales. The success of high-tech greenhouses and vertical farms may be efficiently assessed by combining IoT metrics, RL benchmarks, and thorough analysis. This provides useful information to improve and optimize future deployments.

## 8.7 Future Directions and Opportunities

The rapid advancement of technology, such as IoT, AI, robotics, and bioengineering, is generating new potential for high-tech greenhouses and vertical farms. The primary focus of these advancements is on global concerns such as resource scarcity, environmental sustainability, and food security [17]. The following outlines key possibilities and future directions:

### 8.7.1 Advanced Automation and Robotics

Fully automated systems are revolutionizing agriculture with the arrival of robots that can accurately and effectively carry out tasks such as planting, pruning, harvesting, and packaging. Robotic arms and drones boost productivity while decreasing manual work, making pollination another sector where automation is prominent. Accompanying these capabilities is AI-driven precision farming, which uses self-regulating gear to evaluate the state of crops, maximize the use of resources in real-time, and increase overall yields. Swarm robots are also being utilized for tasks such as monitoring soil health and controlling insects, helping to make farming more sustainable and successful.

### 8.7.2 Integration of Artificial Intelligence and Machine Learning

To take preemptive measures, modern farmers use predictive analytics, which employs AI algorithms to assess crucial variables such as crop yield, fertilizer requirements, and potential disease outbreaks. Machine learning algorithms also dynamically allocate resources like lighting, ventilation, and irrigation based on real-time data, ensuring optimal growing conditions and efficient use of resources. Additionally, edge computing enables farms to make AI-driven decisions in real time, significantly lowering latency and improving the responsiveness of agricultural processes.

### 8.7.3 Renewable Energy and Sustainability Initiatives

Energy-efficient agricultural operations are fueled using renewable energy sources, such as solar panels, wind turbines, and biogas systems, to power buildings sustainably [18]. These are augmented by advanced battery storage solutions that ensure a consistent energy supply even during periods of inadequate generation. To establish a more sustainable and resource-efficient agricultural ecosystem, water conservation is prioritized using closed-loop irrigation systems that significantly minimize water waste, together with advanced recycling and purification systems for hydroponic and aeroponic systems.

### 8.7.4 Multi-Crop and Specialized Farming

In advanced agriculture, crop diversification emphasizes the cultivation of high-value plants such as spices, medicinal herbs, and specialty crops in regulated environments to enhance quality and yield. Customizing facilities for certain crops, such as microgreens, leafy greens, or strawberries, enhances production by ensuring optimal growth conditions. Moreover, aquaponics integration combines hydroponics with aquaculture to establish a closed-loop, sustainable ecosystem that efficiently recycles water and nutrients, sustainiing both aquatic and plant organisms.

### 8.7.5 Vertical Farming in Urban Settings

Urban integration agriculture aims to make full use of available land (green space) in densely populated areas and reusing otherwise barren city infrastructure, i.e., basements and roofs, as productive vertical farms. This model of maximal yield, the production of as much fresh produce as possible, in the area where it can be delivered directly to consumers, supports local production and minimal movement and associated $CO_2$ emissions. In addition, community agriculture projects ensure food security, sustainability, and the urban community's involvement in the urban fabric, since the latter can purchase food through the complementary use of the units of the communal vertical farming and others.

### 8.7.6 Enhanced Lighting and Climate Control Systems

One of the state-of-the-art examples of smart agriculture is the smart lighting systems that employ LED lights, which vigorously alternate their spectrum and intensity to optimally enhance photosynthesis while paying attention to energy conservation. Adaptive climate control aids in further enhancing efficiency by using AI-based systems for self-replacement of temperature, humidity, and $CO_2$ at specific levels depending on the crop. These systems, alongside thermal energy recovery and reuse, help maintain operational temperatures and reduce energy expenditure, contributing to greater control and sustainable management of the agricultural system.

## 8.8 Conclusion

High-tech greenhouses and vertical farms represent a transformative progression in agriculture, offering innovative, efficient, and sustainable solutions to global challenges such as resource shortages, climate change, and food security. Modern agricultural methods provide year-round production with minimal environmental effect by integrating advanced technology such as automation, AI, the IoT, and renewable energy systems. They promote localized food production, particularly in urban environments, reduce reliance on conventional agriculture, and optimize resource utilization. High-tech greenhouses and vertical farms are poised to play a vital role in establishing a resilient and sustainable food system for the future, fostering healthier ecosystems and thriving communities globally as research and development unveil new opportunities.

High-tech greenhouses and vertical farming are idealistic examples of new-age farming techniques that have consolidated various important aspects and promoted sustainability and yield. A variety of mechanical products like automations, robots, AI, and the IoT have changed farming operations through innovative technology enrollments that led to precision farming, real-time control, and resource management. At the same time, such systems also enhance sustainability through closed-loop ecosystems, water recycling, and renewable energy. Through the development of urban agriculture, where urban areas are turned into vertical farms, food production within the urban environment has been made possible through the utilization of vertical farms, which are a means of food security and low carbon footprints. These environments make it possible to grow high-value crops and specialty products that have improved yields and nutritional value due to the controlled growth conditions that enhance crop diversification and improvement. Being systematic in this manner also holds social and economic benefits by advancing community participation and economic development, as well as creating new company models, community agriculture programs, and high-tech employment opportunities.

**AI disclosure statement:** The author confirms that no artificial intelligence (AI)-generated content or AI-assisted writing tools were used in the preparation of this chapter (including summaries and the abstract). All text, analysis, and conclusions are the result of the author's original work.

## References


1. Verma, D., Mishra, A., Mishra, K., Role of IOT in introducing Smart Agriculture. *Int. J. Res. Eng. Technol.*, 12, 9, 2022.
2. Kelly, T.D., Foster, T., David, M., Schultz, D., Assessing the value of deep reinforcement learning for irrigation scheduling. *Smart. Agr. Technol.*, 7, 100403, 2024.
3. Jadon, J. and Singh, R., Challenges and Opportunities of Internet of Things in Smart Agriculture: A Review. *Lect. Notes Electr. Eng.*, 32, 653, 2022.
4. Frommberger, L., Foundations of Reinforcement Learning, in: *Qualitative Spatial Abstraction in Reinforcement Learning. Cognitive Technologies*, Springer, Berlin, Heidelberg, 2010.
5. Sagar, S., Hyder, W., Nisar, A., Shaikh, M.Z., Pabani, J., System Architecture for the Internet of Things Based Smart Agriculture Monitoring. *J. Inf. Commun.*, 16, 13, 2022.
6. Farooq, S., Javid, R., Riaz, S., Atal, Z., Internet of Things (IoT) Technologies in Greenhouse Farming: A Systematic Literature Review. *IEEE*, 12, 23, 2022.
7. Platero-Horcajadas, M., Pardo-Pina, S., Cámara-Zapata, J., Brenes-Carranza, J.A., Ferrández-Pastor, F.J., Enhancing Greenhouse Efficiency: Integrating IoT and Reinforcement Learning for Optimized Climate Control. *Sensors*, 24, 8109, 2024.
8. Ali, S., Mostafa, G.A., Shahidinejad, A., A survey on the computation offloading approaches in mobile edge computing: A machine learning-based perspective. *J. Comput. Netw.*, 182, 107496, 2020.
9. Malabadi, R., Kolkar, K., Chalannavar, R., Castaño, K., Sultan, S., Baijnath, H., Munhoz, A., Abdi, G., Greenhouse farming: Hydroponic vertical farming-Internet of Things (IOT) Technologies: An updated review. *J. Adv. Res. Rev.*, 23, 2634, 2024.
10. Rambod, A., Nastaran, R., Siva, K.B., Arash, B.S., Hazandy, A.H., Application of digital technologies for ensuring agricultural productivity. *Heliyon*, 9, 12, 2023.
11. Jing, H., Terry, L., Weiguang, L., Sachin, C., Subbu, S., Mark, G., David, T., Zhong-Hua, C., Energy and fertigation consumption and economic viability of a light blocking film in greenhouse lettuce production. *J. Clean. Prod.*, 479, 144013, 2024.
12. Till, W., Aidong, Y., Michael, W.H., Energy optimisation of plant factories and greenhouses for different climatic conditions. *Energy Convers. Manag.*, 243, 114336, 2021.
13. Orsini, F., Carotti, L., Souri, M.K., Pennisi, G., Gianquinto, G., Optimizing energy and other resource use in vertical farms, in: *Adv. pl. fact: New tech. in indoor ver. Farming*, vol. 35, p. 125, 2023.
14. Ajit, S.R., Sushabhan, C., Abhinav, S., Pankaj, N., Gautam, S., Empowering vertical farming through IoT and AI-Driven technologies: A comprehensive review. *Heliyon*, 10, 3451, 2024.

15. Gollart, M., Okhrin, Y., A reinforcement learning approach to dynamic portfolio optimization. *Ann. Oper. Res.*, 2025.
16. Vitorino, J., Andrade, R., Praça, I., Sousa, O., Maia, E., A Comparative Analysis of Machine Learning Techniques for IoT Intrusion Detection, in: *Foundations and Practice of Security*, E. Aïmeur, M. Laurent, R. Yaich, B. Dupont, J. Garcia-Alfaro, (Eds.), pp. 191–207, Springer International Publishing, 2022.
17. Tiwari, P.K., Singh, P., Rajagopal, N.K., Deepa, K., Gulavani, S., Verma, A.K., Yekula, P., IoT-Based Reinforcement Learning Using Probabilistic Model for Determining Extensive Exploration through Computational Intelligence for Next-Generation Techniques. *Comput. Intel. Neurosc.*, 32, 5113417, 2023.
18. Karanisa, T., Achour, Y., Ouammi, A., *et al.*, Smart greenhouses as the path towards precision agriculture in the food-energy and water nexus: case study of Qatar. *Environ. Syst. Decis.*, 42, 521, 2022.

# 9

# Case Study on the Initialization of Mapping between the Raw Data and Crop Yield Values for Yield Prediction

**Dharani Jaganathan[1]*, Vishnu Kumar Kaliappan[1], Mani Deepak Choudhry[2] and Sam Goundar[3]**

*[1]Department of Computer Science and Engineering, KPR Institute of Engineering and Technology, Arasur, Coimbatore, India*
*[2]Department of Computing Technologies, SRM Institute of Science and Technology, Kattankulathur, Tamil Nadu, India*
*[3]Department of Computer Science, University of Central Asia, Bishkek, Kyrgyz Republic*

## *Abstract*
Agriculture is an important sector of society, as a significant portion of the world's food supply. However, many countries still face hunger due to food shortages. Predicting crop yields from environmental parameters is key to boosting production and tackling food crises. Deep learning models help yield predictions by extracting features from agricultural data, but face limitations in mapping and feature selection. To enhance the initialization process of mapping raw agricultural data to crop yield values, this chapter introduces a hybrid approach, combining deep learning with reinforcement learning (RL), an important aspect of yield prediction. Unlike conventional deep learning approaches, which struggle with feature selection and spatial-temporal correlations, RL dynamically refines model parameters to enhance predictive accuracy and improve the convergence rate of yield estimations. Through extensive validation, the study demonstrates the superiority of RL in handling heterogeneous agricultural data, leading to more precise and adaptive yield predictions. This chapter shows, through thorough testing, that the proposed model is better at managing different types of agricultural data, resulting in more accurate and flexible yield predictions.

**Corresponding author*: dharanijagan92@gmail.com; ORCID: 0000-0003-1781-7618

Anitha Velu, Prasanth Aruchamy, Raghu Ramamoorthy, Rajesh Kumar Dhanaraj and Seifedine Kadry (eds.) Climate Smart Agriculture, (173–194) © 2026 Scrivener Publishing LLC

***Keywords*:** Crop yield, mapping, precision agriculture, feature mapping, decision-making, crop production

## 9.1 Role of Crop Yield Management Systems

Agriculture plays a vital role in tackling global challenges, particularly ensuring food security, as the world's population is expected to reach 9.7 billion by 2050 [1]. Concentrating crop production is essential because of the present population expansion and global climate change. Crop yield management is a cornerstone of advancing agricultural practices, improving food security, optimizing natural resource use, and promoting sustainable farming.

Effective crop yield management allows farmers to make better decisions regarding resource allocation, planting schedules, pest control, and market timing. This leads to higher productivity while minimizing environmental impact. By optimizing every phase of the crop production cycle from planting to harvesting, farmers can enhance returns, reduce financial risks, and contribute to sustainable food production. Additionally, these practices help meet the growing food demand without expanding agricultural land, which is constrained by urbanization, environmental degradation, and limited resources. Simple yet impactful practices can significantly boost yields by planting crops in the appropriate season, using optimal spacing, and applying fertilizers and irrigation at the right stages.

### 9.1.1 Crop Yield Management in Addressing Climate and Environmental Challenges

Agricultural productivity depends on technological, biological, and environmental factors. When these factors are not properly monitored, they create risks to the farmlands, leading to a yield reduction that may result in economic losses and food shortages [2]. Environmental factors can be classified into abiotic and biotic; both are strongly connected to a change in weather conditions [3]. Abiotic factors include soil properties such as pH, fertility of soil, and climate stresses like flood and drought. The biotic factors include beneficial organisms like decomposers, pests, and weeds that can hinder crop growth. Technological innovations like fertilizer technology and genetic improvements have achieved reasonable success in crop productivity, but inappropriate use of fertilizers and early harvesting can lead to adverse effects.

To overcome the factors that cause issues to yield production, crop yield management comes as a solution. To solve pests and seasonal diseases, this method will provide early detection and management support to the farmer, helping to avoid losses. Similarly, recent technologies play a major role in yielding the crop at a high level. Remote sensing monitors the farmland with the help of satellite images, whereas precision agriculture helps optimize the resources of crops and farmland based on real-time data.

Integrating artificial intelligence techniques helps evaluate data, enabling them to track and predict weather patterns and soil moisture content [4]. Also, help the farmer decide the appropriate use of fertilizers and whether pest control is based on analyzed data. Additionally, the data from the sensor can predict and regulate water flow. Wireless sensors and big data analytics support retrieving and managing real-time data. The discussed recent technologies help prevent farmland degradation and protect water quality, improve productivity, use resources more efficiently, and assist farmers in adopting more sustainable agricultural practices.

The amount of raw agricultural data from environmental sensors is large, and many sensors are used, including environmental and satellite weather stations. The raw data retrieved from environmental sensors, satellites, and other stations produces a large volume of data is a difficult challenge because the data may be in any form, complex, or unstructured. Also, there are missing values that may cause an issue in the accuracy and reliability of the process. The effective data preprocessing techniques, like normalization and transformation, are applied to handle these issues by converting raw data into usable formats to achieve better analytical outcomes and improve data quality [5].

## 9.2 Challenges in Handling Raw Agricultural Data

In crop yield management, data plays a major role in supporting decision-making. However, the data collected from various sources are raw, highly diverse, and complex. This section discusses challenges in handling raw agricultural data, including environmental and operational information from various sensors.

### 9.2.1 Data Heterogeneity and Integration

The diversity of data collected from various resources in different formats, structures, resolutions, accuracies, and scales is known as data heterogeneity, which is the most significant challenge in handling agricultural data. The data comes from sensors, drones, weather stations, and manual records, making it challenging to integrate and analyze.

Remote sensing imaging techniques are used to explore the farm based on its interaction with light. Multispectral imaging is used to monitor vegetation health and soil mapping by capturing fewer spectral bands, while another one is hyperspectral imaging provides a continuous spectrum that helps detect diseases in crops and nutrient analysis with finer details. but combining these data into a simple framework may create complications while processing due to differences in data format and resolutions. Also, providing consistency across these diverse sources is difficult, often leading to inaccuracies in analysis and decision-making.

### 9.2.2 Data Quality and Noise

Crop yield management techniques rely on raw agricultural data that comes from the sensors, but some factors may introduce noise and inaccuracy into this data. Environmental interferences, such as unexpected rainfall, high temperatures, and cyclones, may interrupt the sensor's readings [6]. Similarly, nearby electronic equipment like power lines can cause electronic interference and make noise in the data. Sensors malfunction due to hardware failures, wear over time, and poor calibration, sending incorrect readings due to improper maintenance of the sensor. For example, a malfunctioning temperature sensor may record a high reading instead of the actual lower value, leading to an inaccurate climate assessment of crops. Human mistakes are also affecting the data quality. Data consistency was affected because of typos and incorrect unit conversions.

### 9.2.3 Temporal and Seasonal Variability

Agricultural data are constantly changing and are influenced by seasonal cycles and growth stages of crops. For example, in wheat farming, the wheat crops grow during cooler months. If an unexpected heatwave or cold spells occur during the growing stage, it can be recorded as an anomaly during sensor readings. These anomalies can confuse the dataset and reduce the accuracy of the prediction. Furthermore, the seasonal variability leads to a gap during the off-season, delaying forecasting.

### 9.2.4 Timeliness and Real-Time Analysis

Real-time data collection and analysis are very important in crop yield management because a simple delay may lead to resource wastage or loss in crop yield [7]. For example, if a pest attack is not detected at the right time, crops can be damaged, affecting the harvest. Without real-time data, the farmer also struggles to plan the best time for harvest and selling their products, which can affect their profits.

## 9.3 Reinforcement Learning for Crop Yield Prediction

As discussed in the previous section, handling and mapping the raw data with crop yield is most important. To achieve this, we propose two new hybrid models. The deep Q-network is used for mapping and initializing the data, while linear regression and random forest algorithms select the correct feature to get accurate results. The proposed models are named reinforced linear regression feature selector (RLR-FS) and reinforced random forest (RRF).

## 9.4 Key Components of Reinforcement Learning

Basic key components used to create an RL framework are discussed in this section.

In RL, agent (A) is a learner that makes decisions and interacts with its environment. Based on the policy, which is a set of rules, the agent can take actions in the environment to achieve a higher reward since it is a trial-and-error method, unlike a deep learning model, which has pre-trained data [8].

The environment acts as an external system that interacts with the agent, influencing its actions and determining operational conditions. It gives feedback to the agent for their actions in the form of rewards or penalties. State $(s_t)$ is the agent's perception of the environment at the time(t). It encapsulates all relevant information necessary for decision-making. The choice made by the agent while being in state $s_t$ at time t is known as action $(a_t)$. The reward $r_t$, acts as feedback, guiding the agent's learning process. Policy $(\pi)$ defines the agent's behavior by mapping states to action. It expresses the probability of selecting an action $a_t$ given state $s_t$, denoted as $\pi(a_t|S_t)$, forming the foundation of reinforcement learning strategies.

The value function $V(s_t)$ predicts the total expected reward an agent will obtain from the state [8], if it continues to follow a particular policy, using a discount factor($\gamma$). The value function is computed as follows:

$$\boldsymbol{V(s_t) = E\left[\sum_{t=0}^{\infty} \gamma^t r_t \mid s_0 = s_t\right]} \tag{9.1}$$

Where, $E$ represent the expected value over all possible future states and rewards, similarly the Q-Function $Q(s_t, a_t)$ estimates the total reward, which is computed as follows [9]:

$$Q(s_t, a_t) = E\left[\sum_{t=0}^{\infty} \gamma^t r_t \mid s_0 = s_t, a_0 = a_t\right] \tag{9.2}$$

## 9.5 Reinforcement Learning Algorithms

The following section discusses some of the reinforcement learning-based algorithms.

### 9.5.1 Value-Based Methods (Q-Learning)

Q-learning is a familiar RL algorithm that aims to learn the optimal action-value function $Q*(s, a)$. Equation (9.3) shows the Q-learning update rule [10]:

$$Q(s_t, a_t) \leftarrow Q(s_t, a_t) + \alpha[r_{t+1} + \gamma max_{a'}\ Q(s_{t+1}, a') - Q(s_t, a_t)] \tag{9.3}$$

From Equation 9.3, the Q-learning algorithm updates the action-value estimate $Q(s_t, a_t)$ by incorporating the reward $(r_{t+1})$ obtained and the anticipated future rewards. It adjusts values using a learning rate, discount factor, and the maximum future reward $(\gamma max_{a'}\ Q(s_{t+1}, a'))$ from the next state $(s_{t+1})$.

### 9.5.2 Policy-Based Methods (REINFORCE Algorithm)

Policy-based method REINFORCE is an RL algorithm, in which the agent learns the policy ($\pi$) directly, optimizing it by computing gradients with respect to the expected rewards. This approach maps states to actions to improve decision-making over time, as illustrated in Figure 9.1 [11]. Before updating the policy network $\pi(s, a)$, the objective function $J(\theta)$,

**Step 1 : Initialize the policy function with random weights**
**Step 2 : for each episode (from 1 to K )**
**Set the initial state**
Generate an episode by following the policy, which produces a trajectory consisting of states, actions, rewards:
**Step 3 : For each time step t (from 0 to T) carry out the following steps:**
**Compute the return**

$$C_t \leftarrow \sum_{i=t}^{T} r_i * \gamma^{i-t}$$

Update the policy parameters:

$$\theta \leftarrow \theta + \alpha * G_t * \nabla_\theta log\pi_\theta(s,a)$$

**Figure 9.1** Reinforce algorithm.

defined as the expected sum of rewards, must be evaluated to assess the policy as shown in Equation 9.4.

$$J(\theta) = E_{\pi\theta}[\Sigma_t r_t] \tag{9.4}$$

### 9.5.3 Actor-Critic

This model works by combining a policy and a value function to select actions. The actor $(\theta_{actor})$ learns the policy $(\pi_\theta(s_t))$, which denotes the possibility of taking action $(a_t)$ given state $(s_t)$, with learning rate $(\alpha)$. The actor controls how much the parameters are updated. Meanwhile, the critic evaluates the performance by estimating the state value [8]. Figure 9.2 shows the workflow of the Actor-Critic model [12].

The update rule for the actor and critic is given by:

$$\theta_{actor} \leftarrow \theta_{actor} + \alpha \nabla \log\pi_\theta(s_t).\, \delta_t \tag{9.5}$$

Where the term denotes the gradient $\nabla log\pi_\theta(s_t)$ represents changes in the log probability of actor-selected actions relative to its parameters that lead to policy enhancement. Also, the advantage function $(\delta_t)$ helps to represent the relative value of taking action $a_t$ in state $s_t$ shown in Equation 9.6:

$$\delta_t = r_t + \gamma V(s_{t+1}) - V(s_t) \tag{9.6}$$

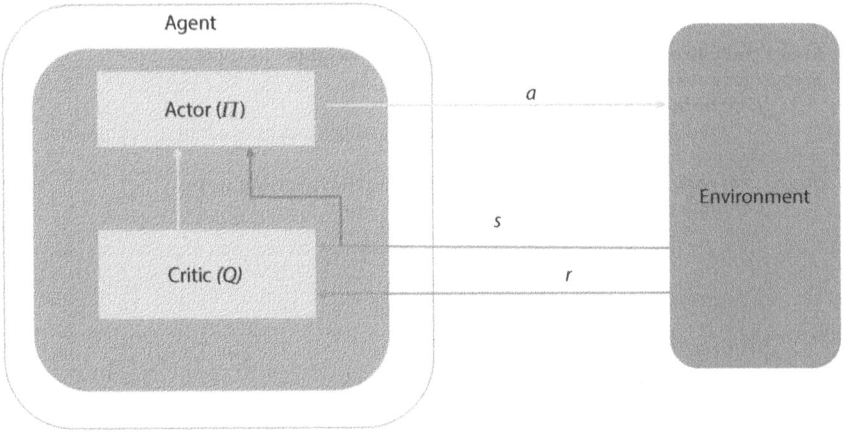


**Figure 9.2** Workflow of the Actor-Critic model.

Similarly, the critic update $(\theta_{critic})$ to estimate the value function is computed as:

$$\theta_{critic} \leftarrow \theta_{critic} + \beta \nabla V(s_t) - r_t \tag{9.7}$$

Here are the critics, learning rate $(\beta)$ controlling how much the parameters are updated.

$\nabla V(s_t) - r_t$: Gradient of the error between the estimated value $V(s_t)$ and the observed reward $r_t$. This adjusts the critic to provide better value estimates.

### 9.5.4 How is Reinforcement Learning Suited for Crop Yield Data Mapping

In agriculture, many factors influencing crop yields, such as soil properties, weather conditions, and irrigation schedules, interact dynamically and in a complex manner. By using RL, crop yield predictions can be refined continuously as more data is collected, ensuring the system adapts to changing conditions such as weather patterns and soil health. Additionally, RL helps integrate both environmental and crop-related parameters into the decision-making processes.

To optimize a pesticide application, a robust RL model is developed by combining with Bayesian modeling in an apple orchard [13]. Threshold policies are applied to control the pest population in various scenarios based on best, worst, and average cases. In results for simple models, the model achieves high rewards while frequently using the pesticide. But for a real-world scenario, it may cause soil damage and possibly irreversible crop damage. Similarly, a crop classification system is proposed with a deep RL model to help farmers select suitable crops for their soil [14]. Deep reinforcement learning (DRL) outperforms crop choices by continuous learning through iterative decision-making when compared with the machine learning model. But DRL outperforms by learning optimal crop choices through iterative decision-making, enhancing crop yield.

Modified Q-learning architecture for UAV path selection, coupled with a U-Net-based convolutional neural network architecture proposed for extrapolating health maps from limited data, and to enable efficient crop health assessment [15]. This proposed system achieves 90% accuracy in generating crop health maps by sampling only 40% of the field, significantly reducing the need for exhaustive scouting. The convolutional neural network-based extrapolation ensures that even with limited data, the generated health maps provide actionable insights for precision agriculture. Similarly, optimizing nitrogen fertilization and irrigation can significantly improve crop management. This is developed as a framework, the decision support system for agrotechnology transfer, for validation by integrating RL and imitation learning [15]. In this, the trained RL policies increased economic profits by up to 45% in Florida and 55% in Zaragoza compared to baseline methods, while also reducing nitrate leaching to mitigate environmental impact. Imitation learning-trained policies, designed for real-world deployment, achieved near-identical performance to RL policies under full observation, with negligible drops in efficiency.

The soil and water assessment tool (SWAT) with RL framework is utilized to optimize crop management strategies to autonomously learn the best actions for crop growth while considering spatial and temporal variations in resource constraints, such as fertilizer and water usage [16]. The performance of RL agents was evaluated in the SWATGym environment using an average reward as the key metric, reflecting the agents' success in optimizing crop yield while minimizing resource inputs like fertilizer and irrigation.

Both DRL and recurrent neural networks are used in learning-based agricultural management to optimize nitrogen fertilization strategies in partially observable and variable climate conditions [17]. The gated recurrent units in Q-networks are used to process sequential observations for agricultural management.

## 9.6 Deep Q-Network

Deep Q-network is a value-based approach introduced in the 2013 NIPS paper Playing Atari and extended in the 2015 paper Human-Level Control through DRL models to determine optimal actions [18]. The Q-function helps the agent to make decisions by estimating the expected future rewards [19].

The following are the key components of the DQN architecture:

- Input Layer: Receives the state representation, which includes environmental and crop-related data.
- Hidden Layers: Multiple layers of neurons process the input to learn complex models.
- Output Layer: This layer is responsible for producing Q values based on the actions taken by the current state.
- Experience Replay: This buffer in DQN stores the current and previous states, experiences, and their associated actions with rewards. It enables random sampling during training, reducing correlations among consecutive experiences.
- Target Network: A separate network enhances learning stability by supplying target Q-values during updates. The architectural diagram of DQN is shown in Figure 9.3.

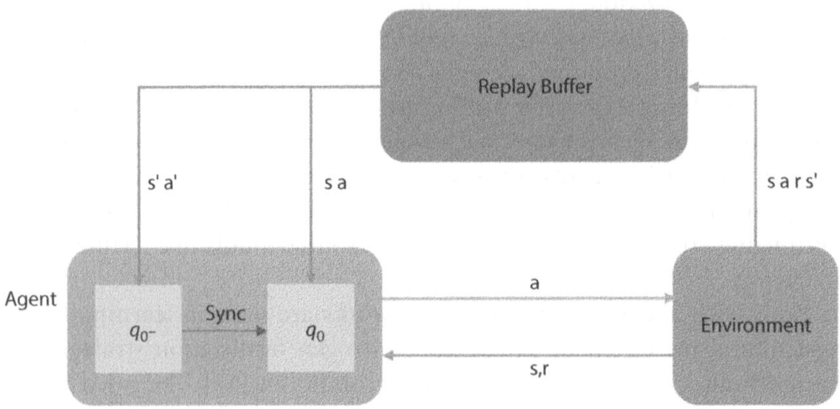


**Figure 9.3** Basic architecture of DQN.

## 9.7 Deep Q-Network Algorithm

Deep Q-network utilizes the Bellman equation to minimize differences between Q-values from preceding and present states when building the optimal action-value function [8].

The present Q-value is used as a target since it is closer to the optimal value. The value of any state-action pair gets calculated by the current reward along with the optimal Q-value available to the subsequent state [20]. Equation 9.8 expresses this relationship

$$q_*(s_t, a_t) = E_{s'}[R_{t+1} + q_*(s_{t+1}, a')] \tag{9.8}$$

### 9.7.1 Q-Learning Update Rule

Through the update mechanism, the estimated Q-value approaches the target Q value at a controlled rate $(R + q(s', a'))$. This is achieved by iteratively refining Q-values to approximate the optimal Q-function, as defined by:

$$q(s,a) \leftarrow q(s,a) + \alpha(R + q(s',a') - q(s,a)) \tag{9.9}$$

### 9.7.2 Integration of Neural Networks

Deep Q-network Q-values get stored inside an approximated neural network that applies parameters (θ) to represent Q-functions. The neural network consumes the state data as input to produce Q-value outputs for all reachable actions according to Equation 9.10. This method allows DQN to scale to complex environments with high-dimensional states, such as images, where a Q-table would be impractical.

$$q_\theta(s,a) \tag{9.10}$$

### 9.7.3 Loss Function

The loss function is designed in DQN to train the neural network that measures the difference between the Q values predicted by the online network $q_\theta(s,a)$ and the target Q-value computed by a separate target network $q_\theta(s',a')$, which is computed as follows:

$$L(\theta) = E[(R + q_{\theta}(s',a') - q_{\theta}(s,a))^{2}] \tag{9.11}$$

Additionally, the gradient descent update rule is used to adjust the parameters theta of the Q-network to minimize the loss, as shown in Equation 9.12.

$$\theta \leftarrow \theta - \alpha \nabla_{\theta} L(\theta) \tag{9.12}$$

The weighting update process makes predictions of Q-values align more closely with target Q-values.

### 9.7.4 Experience Replay

In the proposed model, DL is an integrated RL model. While DL requires independent data, RL environments generate highly correlated data because consecutive states are closely linked. To address correlated training samples, DQN employs an experience replay buffer. This buffer stores transitions $(s,a,r,s')$ and allows the agent to train on randomly sampled experiences. This randomization technique leads to more stable and efficient learning operations.

### 9.7.5 Target Network

To further stabilize the training process, a target network is introduced to prevent rapid oscillations in the Q-value updates. To ensure model stability during training, the target network's parameters $(\theta^{-})$ are periodically updated by copying them from the main Q-network $\theta$ to maintain the model's stability during training. Equation 9.13 illustrates the target Q-value computation.

$$y = R + \gamma max_{a'} q_{\theta} - (s',a') \tag{9.13}$$

## 9.8 Reinforced Random Forest

In RRF, a random forest is applied with two hidden layers to predict crop yield, and DQN is integrated to dynamically select relevant features by

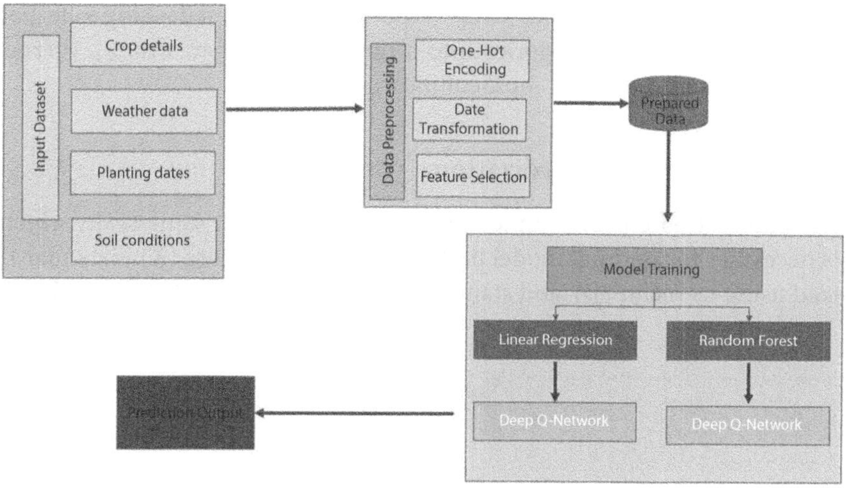


**Figure 9.4** Proposed model architecture.

optimizing rewards. To begin with, the dataset contains details of the crop, planting date, weather conditions, and soil. This dataset helps in forecasting production and influences crop growth. Figure 9.4 illustrates the proposed model architecture.

The features column contains crop type information, crop growth stage, the type of soil required for a particular crop, and insect resistance. To make the raw data compatible, hot encoding is used to convert the categorical values into binary variables. Crop types like corn, wheat, and rice can be represented as binary features for different categories. The crop planting date is converted into a numerical format from a date format to capture temporal significance. This is achieved by converting the planting date into Days Since Planting, which measures the number of days elapsed since a reference date [21].

## 9.8.1 Data Preprocessing

Data preprocessing begins with preparing the dataset by extracting meaningful features. For example, Days Since Planting is calculated to capture temporal relationships with crop yield:

$$Days\ since\ Planting = (Planting\ date - Reference\ Date).\ days \tag{9.14}$$

Categorical features like Crop Type are transformed using one-hot encoding to convert categories into a numerical format. $drop_{first} = True$ argument is used to avoid multicollinearity issues.

### 9.8.2 Feature Standardization

Standardization ensures numerical features are on the same scale, which is necessary for optimal model performance. Each feature (X) is standardized using its mean $(\mu)$ and standard deviation $(\sigma)$, transforming it into:

$$X_{scaled} = \frac{X - \mu}{\sigma}$$

### 9.8.3 Model Training and Validation

The proposed model is evaluated using data points (n), actual crop yield $(y_i)$ and the predicted crop yield $(\widehat{y_i})$. This model is designed with a neural network containing two hidden layers that are trained using the Adam optimizer. The mean squared error loss function serves as a measurement tool that analyzes the differences between predicted output and real target outputs [22]. The performance evaluation uses the root mean squared error (RMSE) approach to create a fair assessment method defined as:

$$RMSE = \sqrt{\frac{1}{n} \sum_{i=1}^{n} (y_i - \widehat{y_i})^2} \quad (9.15)$$

### 9.8.4 Q-Learning for Feature Selection

The state in the DQN framework includes crop-specific details along with other features like planting dates, weather data, and soil characteristics. This enriches the feature set and provides the model with more contextual information for feature selection. Decisions could involve selecting specific crop-related features in combination with environmental or agronomic features to form an optimal subset.

Q-learning dynamically selects relevant features to improve model performance. At each step, the agent decides whether to select or deselect a feature based on its present state, which is shown in Equation 9.18 [23].

$$Q(s,a) \leftarrow Q(s,a) + \alpha[R + \gamma max_{a'} Q(s',a') - Q(s,a)] \quad (9.16)$$

This training integrates random forest with Q-learning to dynamically select optimal feature subsets. The reward for each action is derived from the RMSE of the random forest model trained on the selected features. Over multiple episodes, the agent learns to minimize RMSE by refining the Q-values. Cumulative reward is tracked to measure the progress:

$$\textit{Cumulative Reward} = \sum_{i=1}^{n} reward_i \quad (9.17)$$

Where $reward_i$ is the reward obtained in the $i^{th}$ episode.

### 9.8.5 Reward Tracking and Analysis

The effectiveness of feature selection is evaluated by tracking cumulative rewards across episodes, as shown in Figure 9.5. Increasing cumulative rewards reflects better feature selection strategies. Visualizing these metrics provides insights into the model's learning behavior and the impact of selected features.

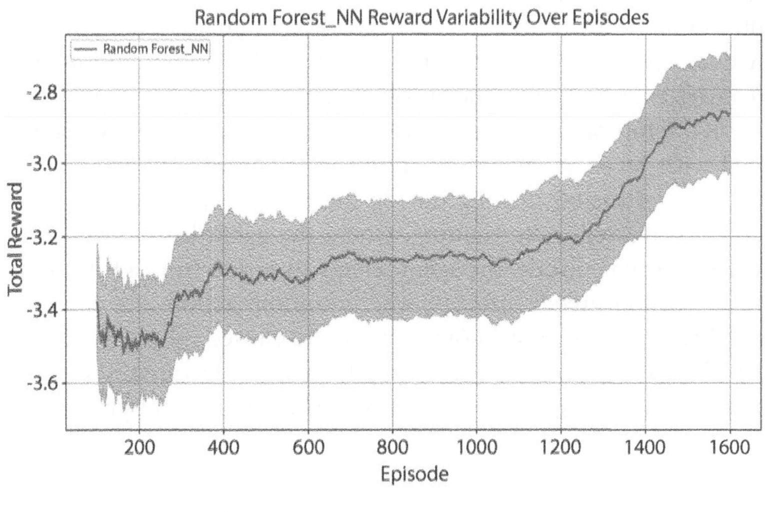


**Figure 9.5** Learning curve for RRF.

## 9.9 Reinforced Linear Regression Feature Selector

The RLR-FS approach combines linear regression and DQN to optimize feature selection for predicting crop yields [24]. Linear regression models the relationship between features and crop yield; at the same time, DQN learns to select the best subset of features by exploring and evaluating their impact on model accuracy using a reward system. Over multiple training episodes, the process improves prediction accuracy by identifying the most relevant features. Linear regression is trained using the training dataset, where the features are related to the target variable, i.e., crop yield.

Data scaling and normalization are essential to ensure that features with different scales do not dominate the model's learning process. Linear regression is sensitive to the scale of input features and can become biased if one feature has a much larger scale than others. To prevent this bias, a normalization technique is applied to ensure that all features have an equal impact on the model. After feature engineering, the data requires splitting into 80–20 proportions to evaluate model performance using training data and testing data. Equation 9.18 describes an RLR model represented as:

$$Y = \beta_O \sum_{i=1}^{n} \beta_i X_i + \varepsilon \tag{9.18}$$

Where $Y$ is $X_i$ represents the $i^{th}$ feature, $\beta_O$ is the intercept, $\beta_i$ is the coefficient of the $i^{th}$ feature. Linear regression evaluates the significance of individual features by estimating their coefficients $(\beta_i)$ with the error term $\varepsilon$. Features with higher coefficients have a greater impact on crop yield.

The RL setup begins with Q-table initialization, where a Q-table, initialized with zeros, is created to store the Q-values. This table serves as the basis for decision-making during the learning process. The training process involves episodic training, where the agent learns over multiple episodes. In each episode, the agent selects an action from a feature subset using an epsilon-greedy strategy, where random actions are chosen with probability $\epsilon$, and the best-known actions are chosen otherwise. The agent trains a linear regression model using the selected features. The reward is computed as the negative RMSE between the original and predicted crop yield values.

### 9.9.1 Reward Tracking and Analysis

During training, the cumulative reward for each episode is tracked to monitor the agent's progress by visualizing the learning curve, which is

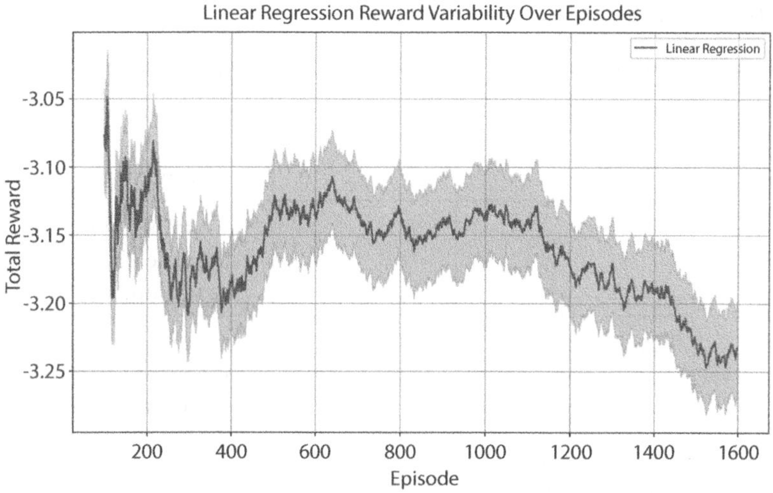


**Figure 9.6** Learning curve for RLR-FS.

shown in Figure 9.6. This reward provides insights into how effectively the agent is learning to select optimal feature subsets.

## 9.10 Experimental Setup and Parameter Optimization

The dataset comprises 1000 rows of synthetically generated agricultural data, encompassing factors of soil characteristics, such as pH levels, organic matter content, weather conditions, and crop yield data like historical yield records, and growth stages. This diverse dataset ensures the representation of the non-linear and dynamic relationships between soil health, weather patterns, and crop yields, offering a realistic test bed for the models. Table 9.1 shows the optimal parameters, such as learning rates, batch sizes, and number of trees, which were determined using grid search.

## 9.11 Results and Discussion

The learning curves of RLR-FS and RRF models reveal their training dynamics and performance trends, as illustrated in Figure 9.7. The RLR-FS model initially shows a rapid decrease in training loss but plateaus early,

**Table 9.1** Hyperparameters configuration.

| Model | Parameter | Range | Hyperparameter detail |
|---|---|---|---|
| RLR-FS | Learning Rate | 0.001 | Step size for weight updates in gradient descent |
| | Batch Size | 128 | The number of sample data used before the model updates. |
| | Epochs | 1600 | The number of times the model processes each sample in the training dataset |
| | Training-Validation (in %) | 80–20% | Proportion of data allocated for training and validation. |
| RRF | Number of Trees | 100 | Number of decision trees |
| | Maximum Depth | 10 | Maximum depth of each tree. |
| | Feature Importance Weight | Adjusted dynamically | Feature weights are updated using Q-learning during training. |
| | Cross-Validation | k = 10 | Technique for assessing model robustness by splitting data into k subsets. |

indicating sensitivity to overfitting and poor generalization. But in RRF, the learning curves are stable and increasing gradually.

Based on the results and the comparison graph, the proposed RRF model is more suitable for handling complex raw data.

## 9.12 Conclusion and Future Scope

Mapping the data with complex relationships between environmental factors and yields is difficult. This chapter concentrated on dynamically

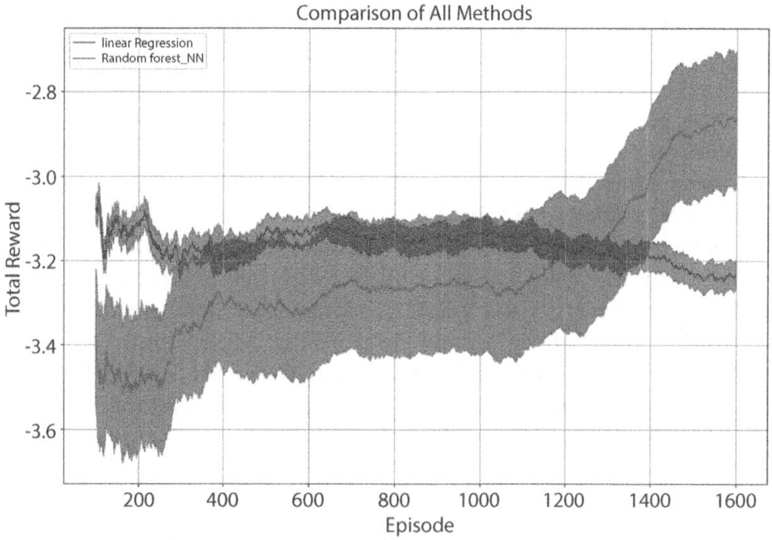


**Figure 9.7** Comparison graph.

addressing these challenges by integrating deep and reinforcement learning techniques. Two novel methods, RLR-FS and RRF, are proposed in this chapter. With the proposed model, the RLR-FS has experienced issues capturing non-linear interactions, as reflected in the learning curve, and may decrease the model's effectiveness. On the other hand, the RRF model achieves high accuracy, but it needs high computational power when applied to large-scale industrial integration, potentially hindering widespread deployment.

Future research could extend this work by integrating additional features, such as planting schedules, to improve predictive accuracy. Also, this chapter primarily concentrated on linear regression and random forest models, exploring hybrid algorithmic approaches to provide more robust solutions for complex relationships. Furthermore, integrating these models with big data analytics and cloud computing could support the development of digital twin-based mapping systems.

**AI disclosure statement:** The author confirms that no artificial intelligence (AI)-generated content or AI-assisted writing tools were used in the preparation of this chapter (including summaries and the abstract). All text, analysis, and conclusions are the result of the author's original work.

## References

1. Silva, G., Feeding the world in 2050 and beyond – Part 1: Productivity challenges, Michigan State University Extension, 3, 1812, 2018.
2. Demircan, M., Gürkan, H., Eskioğlu, O., Arabacı, H., Coşkun, M., Climate Change Projections for Turkey: Three Models and Two Scenarios. *Turk. J. Water Sci. Manag.*, 1, 22, 2017.
3. Liliane, T.N. and Charles, M.S., Factors Affecting Yield of Crops, in: *Agronomy*, A. Amanullah, (Ed.), pp. 112–131, IntechOpen, Rijeka, 2020.
4. Balasubramaniam, S. and Vijesh, C., Computer Vision Systems in Livestock Farming, Poultry Farming, and Fish Farming: Applications, Use Cases, and Research Directions. *Com. Vis. Smart Agri. Crop Man.*, 42, 221, 2025.
5. Suresh, A., Krishnan, P., Jha, G.K., Reddy, A.A., Agricultural Sustainability and Its Trends in India: A Macro-Level Index-Based Empirical Evaluation. *Sustainability*, 14, 2540, 2022.
6. Fizza, K., Jayaraman, P.P., Banerjee, A., Georgakopoulos, D., Ranjan, R., Evaluating sensor data quality in Internet of Things smart agriculture applications. *IEEE Micro*, 42, 51, 2021.
7. Bandaru, L., Irigireddy, B.C., Davis, B., DeepQC: A deep learning system for automatic quality control of *in-situ* soil moisture sensor time series data, *arXiv:2311.06735*, 41, 245, 2023.
8. Jaganathan, D. and Kaliappan, V.K., Optimizing urban sustainability: Reinforcement learning-driven energy-efficient ubiquitous robots for smart cities, in: *Intelligent Solutions for Sustainable Power Grids*, K.L. Ashok, S. Angalaeswari, S.K. Mohana, C.B. Ramesh, P. Arunkumar, (Eds.), pp. 245–271, IGI Global Scientific Publishing, Hershey, PA, USA, 2024.
9. Kröse, B.J.A., Learning from delayed rewards. *Robot. Auton. Syst.*, 15, 46, 1995.
10. Chadi, M.A. and Mousannif, H., Understanding reinforcement learning algorithms: The progress from basic Q-learning to proximal policy optimization, *arXiv:2304.00026*, 8, 457, 2023.
11. Sutton, R.S., McAllester, D., Singh, S., Mansour, Y., Policy gradient methods for reinforcement learning with function approximation. *Adv. Neural Inf. Process Syst.*, 14, 1057, 2000.
12. Konda, V.R. and Tsitsiklis, J.N., On actor-critic algorithms. *SIAM J. Control Optim.*, 42, 4, 2003.
13. Siddique, T., Hau, J.L., Atallah, S., Petrik, M., Robust pest management using reinforcement learning. *Mul-Dis. RL Dec. Mak.*, 41, 231, 2019.
14. Bouni, M., Hssina, B., Douzi, K., Douzi, S., Towards an efficient recommender system in smart agriculture: A deep reinforcement learning approach. *Procedia Comput. Sci.*, 12, 203, 2022.

15. Ning, Y., Li, N., Cheng, J., Wang, B., Peng, Y., Qin, L., A modified Q-learning algorithm for path planning and obstacle avoidance for unmanned aerial vehicles, in: *Proceedings of 3rd 2023 International Conference on Autonomous Unmanned Systems (3rd ICAUS 2023)*, pp. 247–258, 2024.
16. Madondo, M., Azmat, M., Dipietro, K., Horesh, R., Jacobs, M., Bawa, A., Srinivasan, R., O'Donncha, F., A SWAT-based reinforcement learning framework for crop management, *arXiv:2302.04988*, 23, 421, 2023.
17. Wang, Z., Xiao, S., Li, J., Wang, J., Learning-based agricultural management in partially observable environments subject to climate variability, *arXiv:2401.01273*, 24, 273, 2024.
18. Roderick, M., MacGlashan, J., Tellex, S., Implementing the deep Q-network, *arXiv:1711.07484*, 17, 478, 2017.
19. Alkaff, M., Basuhail, A., Sari, Y., Optimizing water use in maize irrigation with reinforcement learning. *Mathematics*, 13, 595, 2025.
20. Wiering, M. and Otterlo, M.V., *Reinforcement Learning: State-of-the-Art*, pp. 1–638, Springer, Berlin, Heidelberg, 2012.
21. Sindhwani, N., Anand, R., George, A., Pandey, D., *Robotics and Automation in Industry 4.0*, pp. 1–326, CRC Press, Boca Raton, 2023.
22. Terven, J., Cordova-Esparza, D.M., Ramirez-Pedraza, A., Chavez-Urbiola, E.A., Romero-Gonzalez, J.A., Loss Functions and Metrics in Deep Learning, *arXiv:2307.02694*, 5, 14, 2023.
23. Kaliappan, V.K., Kuppusamy, M.S., Min, D., *Urban Air Mobility*, pp. 1–345, River Publishers, New York, 2024.
24. Dick, T., Gillenwater, J., Joseph, M., Better private linear regression through better private feature selection, *arXiv:2306.00920, 34, 567*, 2023.

# 10

# Case Study on Reinforcement Learning-Based Decentralized Approach for Precision Agriculture and Environmental Monitoring

**S. Vijayprasath[1*], R. Mohan Raj[1], R. Sathesh Raaj[2] and Ashok Manoharan[3]**

*[1]Department of Electronics and Communication Engineering, PSNA College of Engineering and Technology, Dindigul, Tamil Nadu, India*
*[2]Department of Electronics and Communication Engineering, Vel Tech Rangarajan Dr. Sagunthala R&D Institute of Science and Technology, Avadi, Chennai, India*
*[3]Focuslabs LLC, Dallas, TX, United States of America*

***Abstract***
Precision agriculture and environmental monitoring are necessary to handle the world's resource optimization and sustainability challenges. In this chapter, reinforcement learning (RL) methods are integrated into the development of intelligent systems that can make adaptive decisions in real-time scenarios. The Internet of Things-enabled sensors, cloud-edge computing, and RL algorithms are presented in a combined system architecture to enable effective solutions for agricultural irrigation and environmental monitoring. Deep Q-networks (DQN) and multi-agent RL (MARL) are exploited for algorithmic advances in resource usage, operational efficiency, and environmental impact. The RL approach is used to optimize resource usage, operational efficiency, and environmental impact. The RL-based system in precision agriculture (PA) uses real-time soil moisture, weather, and crop data to develop optimum irrigation schedules that save water and improve yields. A decentralized RL system balances resource consumption to produce accurate and reliable data for decision-making in dynamic environments as part of environmental monitoring. The methodology is demonstrated through case studies, including RL-driven precision drip irrigation in sugarcane farms and RL-based air quality monitoring in urban settings. Through advanced sensor

*Corresponding author: vpr.research@gmail.com; ORCID: orcid.org/0000-0003-0598-4048

Anitha Velu, Prasanth Aruchamy, Raghu Ramamoorthy, Rajesh Kumar Dhanaraj and Seifedine Kadry (eds.) Climate Smart Agriculture, (195–218) © 2026 Scrivener Publishing LLC

networks, scalable cloud platforms, and data-driven policies, these systems deliver water conservation, pollution control, and sustainability. This opens opportunities for expansion while addressing challenges like data quality, computational complexity, and user adoption. Moreover, a decentralized RL system enables a dynamic balance of resource expenditure through accurate monitoring. The methodology's practicality is illustrated by case studies such as RL-driven precision drip irrigation in sugarcane farms and RL-based air quality monitoring in urban settings.

***Keywords:*** Precision agriculture, environmental monitoring, deep Q-networks, multi-agent RL, resource optimization, dynamic decision-making

# 10.1 Introduction

## 10.1.1 Overview of Precision Agriculture

In more relatable terms, precision farming is the application of technology in farming practices. The goal of precision farming is to boost productivity while using natural, mechanical, and human capital more efficiently with a view to reducing the adverse environmental effects of farming. It tries to resolve this challenge by focusing on crop requirements specific to various areas within a field. Instead of a wide-scale application for all plots, farming methods are targeted and customized according to designated plots where they would be most useful. Imagine having a global positioning system (GPS) in your tractor that helps you figure out the specific areas that need to be fertilized or showered with pesticides. Such advanced technology allows farmers to load or distribute quantities of specific products, including seeds, fertilizers, and chemicals, in specific areas rather than throughout the whole field, which wastes chemicals and money [2]. It is undeniable that the Internet of Things (IoT) devices, machine learning (ML), and remote sensing are increasingly integrating with precision agriculture (PA). But even now, the use of drones, GPS satellites, and GIS telemetry has been clearly proven effective in farming [7].

Where it incorporates modern gadgets and artificial intelligence (AI), this unconventional farming method focuses on GPS, GIS, drones, the IoT, and AI as a point of interest. This information is used for various parameters such as soil health, moisture levels, soil nutrients, crop health, and even weather patterns. This abundance of data allows farmers to choose when to plant crops, apply fertilizers, irrigate fields, and harvest crops for the best outcomes in yield and produce quality [9].

Precision agriculture ensures proper usage of resources and assets, helping to conserve water and reduce pollution from controlled over-dependence on chemical fertilizers, pesticides, and other harmful substances. This enables PA to improve soil nutrient and biodiversity content, as well as to lower greenhouse gas emissions and chemical leaching to adjacent habitats, by placing resources such as water, fertilizers, and pesticides precisely when and where necessary only. In addition to the most commercialized farms, PA can serve the purpose for smallholder farmers as well. In conjunction with the increasing penetration of affordable resources and user-friendly applications, deeper farming practices on smaller farms enable profitable growth with long-term sustainability. The three principles that constitute the foundation of precision farming are observations that involve the use of sensors, drones, and satellite imagery to collect high-resolution data on crop, soil, and environmental conditions. The second one is evaluation, which combines data analytics, AI, and ML to analyze crop requirements and outcomes against a set of conventional and novel settings. Following this last one is intervention, which is based on the evaluation, mechanized actions such as targeted irrigation, and precise fertilization, along with pest control targeted according to evaluation, are progressively executed. A wide range of applications has been developed on the other side of PA to optimize each part of the farming process. Another important application is irrigation control, where precision irrigation systems are equipped with soil moisture sensors. Weather variations can be used to apply just the right amount of water for the crops [10].

Reducing water wastage also supports the Allium health. Variable rate technology, as used in PA, also helps farmers apply fertilizers and soil amendments to specific areas where nutrients are needed to increase their soil fertility without unnecessary fertilizer use through excess [4]. Drone-based early detection systems and machine learning algorithms can be used in pest and disease management to identify signs of pest infestations or disease regarding targeted interventions [14].

Precision agriculture brings several benefits. It will also reduce input costs, such as water, fertilizer, and pesticides, while increasing yields and hence improving the profitability of farming operations [3]. Precision use of resources also has the environmental benefit of reducing chemical run-off and conserving natural resources, that is, contributing to sustainability [15]. Precision agriculture technologies help farmers adapt to climate change to feel weather and soil conditions change in real time and adapt even to changing climate patterns [1]. Nevertheless, several challenges lie in the way of PA. There are high initial costs, involving significant investment, for PA technologies, including drones and automated equipment [5]. Farmers also lack both the skills and knowledge to read and use complex data and to operate advanced systems. Additionally, connectivity problems occur; rural areas often lack the strong internet connection required for IoT and cloud-based solutions [7].

### 10.1.2 Environmental Monitoring in Agriculture

Environmental sustainability is managed by monitoring the coupled agricultural productivity and environmental balance. Agroecological impacts of modern agricultural practices may include soil degradation, water pollution, and biodiversity loss. Farmers can minimize these impacts by continuously monitoring environmental parameters [13]. Soil monitoring, where properties of pH, organic matter content, and moisture levels are measured, uses sensors to ensure soil health. Water quality monitoring with devices tracks parameters that prevent contamination and overuse, like salinity, turbidity, and nutrient levels in irrigation water. Climate monitoring with temperature, rainfall, humidity, and wind speed is collected by weather stations and satellite systems, providing farmers the ability to predict and adjust to weather changes [7]. Crop monitoring where images acquired by drones and cameras in multispectral formations are processed to determine crop stress factors (nutrient deficiencies, pest infestations, etc.).

Environmental monitoring enables sustainable agriculture by promoting resource efficiency, where water, fertilizers, and pesticides are subjects of monitoring tools that can minimize waste and pollution [15]. Soil health should be enhanced, so soil conditions can be regularly assessed through practices that prevent erosion, compaction, and loss of fertility [13]. It also supports biodiversity by monitoring ecosystems and reducing the adverse effects of farming on flora and fauna.

#### *10.1.2.1 Monitoring Technologies for the Environment*

Remote sensing is where satellites and drones capture high-resolution images of fields for what they discover: crop health, soil conditions, and water use [7]. The IoT sensors are used with internet-connected devices to make real-time data from environmental factors such as soil moisture and air temperature available. Both AI and ML algorithms use large datasets to support better decision-making, and patterns and predictions are derived from [14]. There are certain challenges while implementing this technology, and the first is data overload, where farmers often find themselves clutching at straws when it comes to managing and making sense of the sheer volume of monitoring data they need to. The second is the cost of equipment, since high-tech tools available to smallholder farmers are prohibitively expensive. The last is a lack of integration: Many monitoring systems that exist today run in silos, and the systems' outputs are much less like a holistic picture of the farm.

### 10.1.3 Role of Reinforcement Learning

Reinforcement learning is a type of machine learning where an agent learns to make decisions by interacting with its environment and receiving feedback in the form of rewards or penalties [8]. There are certain advantages of RL in agriculture, including adaptability since RL agents can adjust to changing environmental conditions, such as shifts in weather or pest activity [6]. Efficiency is achieved by optimizing resource use; RL reduces costs and minimizes waste [12]. Decentralization, where RL enables individual agents, such as drones or sensors, allows them to operate independently while contributing to a larger system [6]. Some examples of RL applications include irrigation systems where RL algorithms learn to optimize water delivery based on soil moisture and weather forecasts, reducing water waste [10]. Autonomous vehicles with RL-powered drones and tractors perform tasks such as planting, harvesting, and monitoring with minimal human intervention. Pest management, where RL models identify and implement the most effective pest control strategies, balancing crop protection with environmental considerations [12].

### 10.1.4 Objective of the Case Study

This case study aims to demonstrate the potential of reinforcement learning-based decentralized systems for PA and environmental monitoring. The objectives are as follows: This study aims to evaluate the feasibility of applying RL-based systems in the agricultural context, and in particular, whether RL-based systems can achieve optimum use of resources and improved productivity. In addition, the study will aquantitatively determine the gains and losses of using RL-based systems using performance indicators, such as crop yield, water use, and input costs to measure both economic and environmental benefits. The study will also choose and document the impediments that arise while installing these systems, as well as technological, economic, and operational barriers. The secondary objectives are to showcase how RL can reduce waste, conserve resources, and protect ecosystems, thereby promoting sustainable practices.

Finally, the study seeks to contribute valuable insights and recommendations for future research by researchers and developers on AI-driven solutions in the agricultural sector.

## 10.2 Fundamentals of Reinforcement Learning-Based Decentralized Systems

### 10.2.1 Overview of Reinforcement Learning

Reinforcement learning is a newer dimension to AI that lets an agent operate within an environment and make decisions. An agent is supposed to learn based on the trials it undertakes and through a system of rewards and penalties (Figure 10.1). Unlike supervised learning, RL does not require labeled datasets, such as agriculture, where conditions change frequently and dynamic environments [16]. There are several key components in RL related to being optimized. The agent is an individual who must make a choice based on the system in which he or she operates at the present state [11]. The environment is everything outside the agent that it wishes to change or optimize. The state refers to the present condition of the environment, including specific factors like soil moisture or temperature. The agent's actions are triggered by what it observes in the state, like setting irrigation levels. The environment then sends a reward signal that tells the agent what it should learn. A rule or strategy applying to how the agent chooses actions when trying to achieve a long-term reward. It also provides value function estimates with respect to the potential benefit of a state or

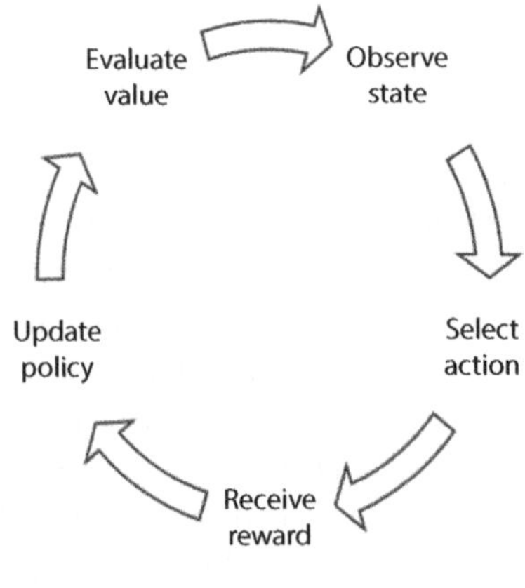


**Figure 10.1** Reinforcement learning cycle.

action in terms of achieving future goals for the agent to not only make better short-term decisions but also long-term success [17]. For instance, in PA, an RL agent would start by watering crops too much or too little, but quickly learns the right amount based on the observation of the crop's health and soil conditions. Moreover, RL systems are robust due to their capacity to incorporate unknown weather conditions, among others [18].

### 10.2.2 Decentralized Systems in Precision Agriculture

Decentralized systems distribute decision-making across multiple nodes or agents, enabling local autonomy and reducing reliance on centralized control [19]. These systems are especially useful in PA, where variations in soil, crop types, and microclimates require localized solutions. Advantages of decentralized systems include scalability suitable for large farms. Resilience is where, in case of failure of one node, the other nodes remain functional. Localized optimization is done using sensors, irrigation, fertilization, and pest control can be adjusted in real time [20]. For example, decentralized systems can optimize water use by enabling sensors across different zones of a field to make independent decisions about irrigation needs. It reduces water wastage, as well as efficiently distributes resources [21].

#### *10.2.2.1 Integration of Reinforcement Learning in Decentralized Agricultural Systems*

Reinforcement learning offers the opportunity for agents, whether drone or sensor, to learn behaviors that can be adopted in resource-sharing most effectively and with the least detrimental environmental consequences [22]. For instance, in pest management, powered by RL algorithms in drones, the former scans the crop's health and relays pest density zones. The drones then determine the best approach for applying the pesticide to avoid the use of chemicals, which keeps the ecosystem diverse [23]. A number of challenges still exist regarding the integration of RL into agricultural systems. In addition, several high computational costs in training RL models can be very resource-intensive and time-consuming. Reinforcement learning systems are hard to integrate into existing agricultural infrastructure: existing systems may not be compatible with them, or they would need to be changed significantly. Also, RL models depend on high-quality data from IoT sensors to make the right decisions. Such receivers should have high accuracy about the inputs received from them [24]. However, the main challenge of making PA more efficient and sustainable has been addressed by RL, as shown in the fact that it has held huge promise in undermining the inefficiency by using different techniques to make the practices more sustainable while still utilizing adequate resources [25]. Typically, an RL framework in PA has several core components. The system uses input data such as soil moisture, temperature, crop type, and weather forecasts, as these are important deciding factors for the system. This data is processed by the RL agent to make important decisions, such as determining the optimal irrigation timing. After these actions, the reward system renders feedback on the outcome of the actions based on factors such as crop yield or resource savings [26, 27]. An RL-based application for an irrigation system that monitors real-time soil moisture and weather data is an example of RL in agriculture. Such a system learned to schedule water delivery to optimize water consumption by 30%, maintaining production at its level [28]. To reduce pesticide use by 35%, RL algorithms were used to deploy drones that would find and assess pest populations only in the infected areas, and apply pesticides [29]. This paper also reports that multi-agent RL (MARL) systems are employed to analyze multispectral image crops where stress is detected and interventions are suggested to improve crop quality and reduce human labor expense [30].

The great potential of RL in PA is hindered by some obstacles, including those encountered in decentralized agricultural systems. Data quality is a major issue. Unless this data is complete or at least not noisy, RL models will be affected, and their accuracy will decrease [33]. For example, the very high setup costs of RL systems can make it difficult to use them in developing countries, as is the case with the economic constraints [34]. Additionally, there is a barrier in the lack of technical expertise of many farmers because they may not be knowledgeable enough to use these advanced technologies properly [35].

### 10.2.3 Environmental Monitoring Using Reinforcement Learning

It is very important to track and assess its functions and impacts when it comes to climate change issues, loss of biological diversity, pollution, and the exhaustion of natural resources. Classical structures for monitoring the environment might be centralized, in this context referring to a single control room gathering data from sensors and making decisions based on that information. Although comprehensively capable of providing solutions for settings, centralized approaches suffer from multiple difficulties: scalability, latency, concentrations of weakness, and flexibility to cover varied local conditions. These challenges are solved by a decentralized RL approach that optimizes general targets across agents and plans their decisions independently. There are certain limitations, like real-time monitoring requires rapid decision-making, which is often delayed in centralized systems due to the time required for data transmission and processing. Environmental monitoring often occurs in remote or harsh conditions where maintaining a central server can be challenging.

Applying decentralized RL, however, provides a robust alternative to centralized systems by distributing the decision-making capabilities between autonomous agents. The system is adaptive, resilient, and efficient as these agents collaborate to achieve a global objective, but act locally. In decentralized RL, agents are independently learning from their local environment without entering a central server. Hierarchical organization of agents permits scalable, multi-agent collaboration over large monitoring networks. New sensors or monitoring units can be added easily because new agents can learn and learn to work by themselves and integrate into the system. The absence of a central control point eliminates single points of failure. However, if a single agent fails, others can remain online. Agents communicate *via* peer-to-peer networks, so they remain active even if some communication links are disrupted [31, 32].

## 10.3 System Design

In this chapter, the system design and integration methodology of RL in PA for optimal decision-making is introduced. The system architecture, data collection methodology, and training the RL model for effective irrigation management are divided into sections.

### 10.3.1 System Architecture

The system design uses RL techniques and is based on an intelligent decision-making platform to control agricultural operations. It contains hardware components, software layers, and engages with data sources for real-time analysis. For monitoring of essential environmental parameters for agriculture, a large number of sensors are deployed. These are soil moisture sensors, weather monitoring stations, and remote sensing. These sensors take data on the level of moisture in the soil, temperature, humidity, and precipitation. Data from these sensors is sent into the central system for processing. The data is processed using cloud computing platforms or on-premise servers that integrate the data from multiple sensors.

Data storage is facilitated by scalable cloud-based storage systems to ensure that large volumes of agricultural data can be handled efficiently. The decision-making is handled by an intelligent model developed with RL algorithms, which performs real-time processing of the input sensor data and sends commands to the irrigation systems. The irrigation system is automated to adjust according to the real-time recommendations of the RL agent. For farmers and agricultural managers, a graphical user interface that they can use to interact with the system. The interface presents real-time information, historical trends of the data, and suggests improvements based on model recommendations.

### 10.3.2 Reinforcement Learning in Agriculture

The system methodology section explores the ways in which the topic of irrigation management applies to RL. By making successive decisions based on the RL application, this system is adaptive to changes in its environment. The primary function of the system will be optimizing irrigation schedules such that the soil moisture levels are within certain acceptable limits while using minimal amounts of water. In RL, this problem can be formulated with a Markov decision process, which consists of state (S) characterization parameterization of the system supplied as a vector of environmental factors (soil moisture, weather forecast, crop water requirements, etc.), and a set of actions, which can be taken by RL and Reward (R) defined based on the efficiency of the irrigation system. In this study, the RL model is based on DQN, where Q-values are estimated using deep neural networks. Due to its ability to learn complex decision-making policies in high-dimensional environments, DQN has been widely used in many applications. Deep Q-networks, when scaling the Q-learning algorithm, a neural network is used that receives the state as input and produces Q-values for each action as output. < Q-values for current state and each action (expected future reward, in terms of [0, 1]) DQN is trained based on the Bellman equation that updates the Q-values in accordance with the rewards received after performing actions.

### 10.3.3 Implementation of the Reinforcement Learning Model in Agriculture

Sensor data is then cleaned and normalized to stay consistent to prepare it for model training. To learn temporal dependencies in the time-series data, such as weather conditions and moisture levels, the deep Q network is made to start with uninitialized random weights and trained in multiple episodes. A reward function is designed to measure irrigation goals, and higher rewards for optimal moisture levels, but low water usage. Considering the outcome of the irrigation actions performed by the agent interacting with the environment, each episode is a cycle of decision-making. The overall process of RL modeling implementation is illustrated in Figure 10.2. System failure to keep soil moisture at required levels is rewarded with a negative reward, and water saved from being wasted is also rewarded with a negative reward. The trained model is evaluated with validation datasets to see how it generalizes to unseen conditions. Model success is measured by performance metrics including cumulative rewards, water usage efficiency, and accuracy in moisture-level predictions. The communication protocol through which an RL agent interacts with IoT devices like water pumps and trivial valves is MQTT. This allows control based on real-time representation fed by agent recommendations. Environmental changes are monitored continuously, and the RL agent decides where to adjust irrigation settings each time step. By its structural features, the system accommodates various crop types and environmental conditions and enables maintaining the most efficient irrigation schedule.

To assess the RL model performance, several metrics are employed. The metric of choice for success is how much water can be saved while keeping soil moisture levels in the healthy range. The objective is to reduce water use and maintain crop yield. This falls under water usage efficiency. Maintaining the soil moisture range at an optimal level is crucial for crop health for calculating soil moisture accuracy. This metric suggests that the system is well capable of maintaining the existing set points when time is considered. The last metric is training time since both speed costs, training the RL model to an optimal policy, and speed costs in real-world deployment, are important. The faster the convergence to an optimal policy, the quicker the RL model will be ready for deployment in the real world. The challenge is data quality analysis, since the accuracy of the model depends

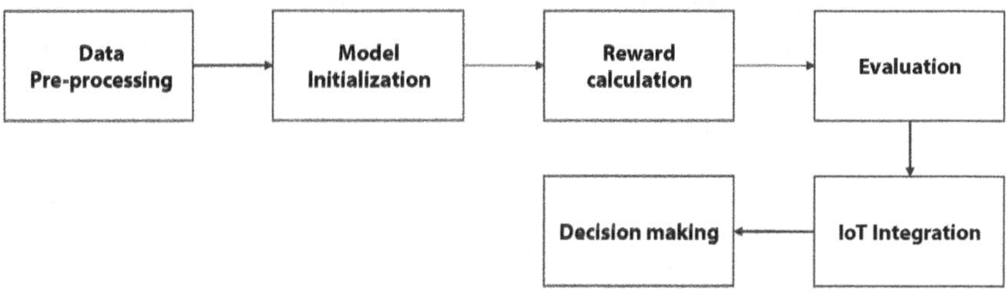


**Figure 10.2** Steps involved in RL model implementation.

critically on the quality of the sensor data. Incorrect decisions can be worsened by poor-quality data or sensor malfunctions. Even if the ability to train these DQNs is desirable, training can be computationally expensive for large datasets or when the model must run in real time, where computational complexity plays the key role. More research is needed to scale the system to bigger agricultural fields or more diverse crop types and farmer adoption, since the system can make optimal decisions, but farmer adoption is a great challenge. For the system to be widely accepted, the user interface must be intuitive, and farmers must have to trust the system.

### 10.3.4 Reinforcement Learning Decentralized Design in Environmental Monitoring

This RL-decentralized system has several interconnected components like sensors, communication networks, RL families, and a cloud-edge computing system (Figure 10.3). The design framework incorporates the following layers, in which the sensing layer collects real-time environmental data such as temperature, humidity, air quality, soil moisture, and water quality. The IoT-enabled sensor is used with preprocessed data. Here, each sensor separately pre-processes and analyzes the local data and only shares it with the network. The sensors are autonomous and preprocess and analyze local data before sharing it with the network. The communication layer features a bilateral data exchange and coordination between decentralized nodes. Computation layers collect data, process, train RL models, and execute decision-making algorithms. Solutions for combining local autonomy with global coordination use edge and cloud computing in a balanced way. The local tasks for distributed computing nodes and collaborative learning mechanisms for global objectives are synchronized. An RL-decentralized system for environmental monitoring consists of multiple interconnected components, including sensors, communication networks, RL agents, and cloud-edge computing systems. The RL algorithms use Q-learning, which works in discrete state-action spaces for simpler tasks. The DQN combines Q-learning with deep neural networks for complex environments. Policy gradient methods directly optimize policies, allowing one to smooth out the decision-making in continuous action spaces. Some examples are MARL algorithms, multi-agent deep deterministic policy gradient, and QMIX. Communication protocols are of two types: Decentralized communication, enabling fault-tolerant, efficient communication with P2P protocols such as gossip algorithms, and hierarchical communication, which combines local communication running in clusters and periodic synchronization among clusters. Hardware and infrastructure consist of

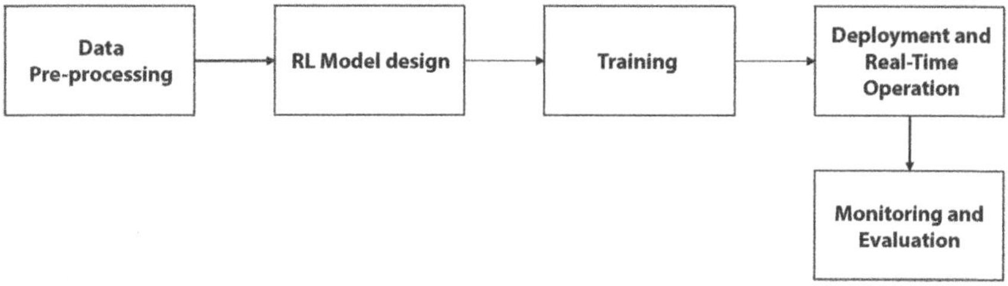


**Figure 10.3** Decentralized RL system for environmental monitoring.

sensor nodes programmable with low-power processors, memory, and communication modules, and edge devices that expose local computation, reduce latency and dependency in cloud servers, enabling massive data analysis and data storage.

This section presents a methodology for designing and implementing a decentralized RL system for environmental monitoring, with a focus on balancing resource utilization while improving monitoring. At first, resource consumption (specifically energy and water) is minimized while capturing the maximum coverage area with minimal data accuracy. It constitutes the state space, that is, variables of air quality, soil conditions, as well as actions such as activating sensors or changing the communication path. Reward functions are cracked in such a way that makes behaviors conducive to low energy use and better anomaly detection incentives. Sensors distributed throughout the monitoring area collect spatial and temporal data types to facilitate data collection. The data must be clean, and the data fed into the RL models should be as clean as possible, where preprocessing steps like normalization, noise removal, and feature extraction are essential. The RL model design includes independent agents, running independently and optimizing local performance concurrently, whilst communicating only for essential updates to minimize overhead. The MARL methods provide the framework in which these agents can collaborate toward global objectives while maintaining local autonomy. With centralized training and decentralized execution, training is done in a simulated environment using global data; execution uses only local information. In a fully decentralized MARL setup, agents can share experiences directly and learn collectively without central coordination. Training occurs within simulated environments that model environmental dynamics and sensor interactions under various resource constraints. Exploration strategies, such as ε-greedy or SoftMax methods, are employed to balance exploration and exploitation during training. Reward-shaping is crucial for guiding agents towards desired policies, emphasizing energy efficiency and accurate anomaly detection. On one hand, RL agents are deployed on edge devices, or sensor nodes, capable of adjusting their policies on the fly as data is received and the environment is changed. Nodes use collaborative learning mechanisms that periodically share their insights, allowing nodes' actions to be coordinated with certain overarching global goals. But continuous monitoring and evaluation are critical for effectiveness under varying conditions. The key performance metrics are accuracy, e.g., how consistently monitoring data matches ground truth measurements, and efficiency, i.e., cost/perspective compared to the amount of environmental data collected. This structured approach to decentralized RL is designed to meet environmental monitoring needs with the optimal use of resources by the centralized RL system.

Certain challenges, like RL, provide the basis for a potentially powerful framework for decentralized approaches, the use of which comes with many challenges and limitations in environmental monitoring. One challenge is the complexity of environmental systems themselves. It is very dynamic and nonlinear. It is also governed by many interdependent variables, like climate patterns, human activities, and the availability of natural resources. This results in a large computational overhead in modeling these interactions exactly in an RL context, which could end up with simplified environments that remove important features of the world. Perhaps the biggest challenge is that the approach is decentralized.

Multiple agents operate autonomously while sharing information in a decentralized system. This means that numerous sensor nodes or agents are spread over a large area, and over vast areas, not to mention in remote and harsh environments, in environmental monitoring. On top of that, data scarcity and quality are other challenges. Reinforcement learning models are very sensitive to the volumes of high-quality data for training and validation. Yet, environmental monitoring suffers from many missing data points through sensor malfunction, poor spatial coverage, and insufficient historical records.

## 10.4 Real-Time Case Studies in the Use of Reinforcement Learning for Precision Agriculture and Environmental Monitoring

In this section, different case studies and applications of RL-decentralized approaches in specific areas are discussed.

### 10.4.1 Case Study: Reinforcement Learning Applied to Precision Drip Irrigation of Sugarcane Farms

Tamil Nadu has the most districts famous for their huge cultivation of sugarcane. The region's climate is semi-arid, and it struggles to manage its water resources. Sugarcane, being a water-intensive crop, has historically been irrigated by flood irrigation practices, causing inefficiency in water use, groundwater depletion, and lower productivity. The case study illustrated in Figure 10.4 looks at the application of RL-based precision drip irrigation systems in sugarcane farms of Erode to address challenges faced in such drip irrigation systems. This technology uses data-driven decision-making to determine the most efficient irrigation schedule, saving water and improving crop yield. The need for precision irrigation is to prevent water wastage, improve nutrient delivery, and support sustainable agricultural practices. Reinforcement learning is a branch of AI where agents learn optimal actions by interacting with their environment. In the context of precision irrigation, RL models can analyze real-time data from sensors (e.g., soil moisture, temperature, humidity), predict the most efficient irrigation schedules, and adjust to changing environmental conditions and crop water requirements. Precision drip irrigation based on RL implementation involves implementing a systematic approach that leverages advanced technologies to enable real-time decisions to improve water use and crop yield. The process starts with an infrastructure setup of high-quality drip lines with variable emitters fed across a 10-hectare sugarcane field pilot area. A selection of Decagon EC-5 or SM300 soil moisture sensors is placed at various depths (e.g., 10 cm, 30 cm) to monitor water levels of the root zone. Weather sensors (e.g., such as the Davis Vantage Pro2) record environmental measurements (temperature, humidity, wind speed, and rainfall), and flow meters (e.g., Netafim meters) measure precisely how much water was delivered. This data is collected and transmitted to a centralized cloud platform by the IoT devices for analysis using RL algorithms. Typically, the central controller (Arduino, Raspberry Pi), based on microcontrollers, processes real-time data, and sends irrigated commands to solenoid valves (Hunter DC solenoids).

In the second phase, designing and training the RL model is focused on using a deep Q-learning approach for its ability to handle complex decision-making problems. The objective of RL is to maximize water-use efficiency and increase sugarcane yield. With historical weather patterns, soil properties, and crop water requirements, the model is trained from data to run simulations of differing irrigation scenarios and identify optimal strategies. Care is taken to craft a reward function for penalizing

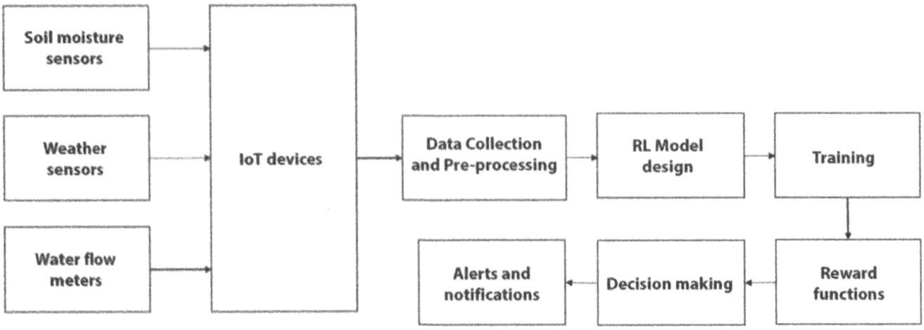


**Figure 10.4** Reinforcement learning-based precision drip irrigation monitoring system.

over- and under-irrigation and rewarding behavior that ensures optimal soil moisture levels, saves water, and also increases yield.

The RL model is in the operational phase: it analyzes in real time incoming data from sensors to predict the most efficient irrigation schedule and optimal water volume to use. For example, if the right soil moisture levels are present, then irrigation is delayed to save water. Solenoid valves receive control signals from the RL model to adjust the water flow as needed. The RL model evolves based on farmers' feedback about crop health and productivity, *via* a feedback loop that tunes the RL model to become more accurate and valuable. With this iterative process, continuous improvement and making our system adaptable to different conditions are achieved. A key interface is a mobile application that provides farmers with an overview of real-time irrigation schedules, water use, and crop performance. For local accessibility, alerts and recommendations are presented in Tamil. The system's backbone is a cloud platform that stores historical and real-time data, executes the RL model, and lets sensors, controllers, and the user interface communicate seamlessly. It is an integrated system that surfaces both how to optimize irrigation and how to turn insights into actionable inputs to contribute to sustainable agriculture and efficient water resources management.

### 10.4.2 Case Study: Banana Farms in a Reinforcement Learning Decentralized Approach

Implementation of the RL-driven water management system for banana farms is focused, as illustrated in Figure 10.5, and the data was collected first using high-tech technology. Across the fields, IoT sensors like Decagon EC-5 and SM300 have been deployed to measure root-zone water levels.

Davis Vantage Pro2 type sensing collects temperature, humidity, rainfall, and wind speed data, while groundwater sensors measure aquifer levels. Communicated to a central cloud platform *via* networks like LoRaWAN or GSM, these devices continuously transmit real-time data to an individual cloud platform. Remote sensing technologies using satellite imagery from Landsat and Sentinel are also used simultaneously at high levels of detail to reveal land use patterns and evapotranspiration rates. Data collected is preprocessed to be accurate and stored in a centralized database for further analysis. Then, a deep Q-learning framework is used to optimize irrigation practices, with the RL model as the optimization. Historical data of weather patterns, soil properties, and crop water instruments are used to train the RL agent that represents the irrigation system. Various irrigation scenarios are simulated for maximizing water use efficiency and crop yield. A reward function penalizes actions that cause over- or under-irrigation, and rewards actions that keep soil moisture at optimal levels. After learning, the RL model uses incoming sensor data to make real-time decisions about irrigation schedules and water volumes. In this case, if soil moisture is adequate, the model delays irrigation to save water. These decisions are integrated into the irrigation system using controls like Arduino or Raspberry Pi, which process commands from the RL model that tell Hunter DC solenoid valves how much water to deliver. Variable flow emitters on high-quality drip irrigation lines deliver water precisely to banana plant root systems, eliminating evaporative losses. Interventions such as mulching, rainwater harvesting, and crop diversification are also automated using the model's guidance. Farm ponds and check dams capture and store rainwater and recharge aquifers when necessary, while organic and plastic mulches reduce soil evaporation. Banana crops are intercropped with legumes in regions where there is severe stress on groundwater for purposes of soil health and water efficiency improvement.

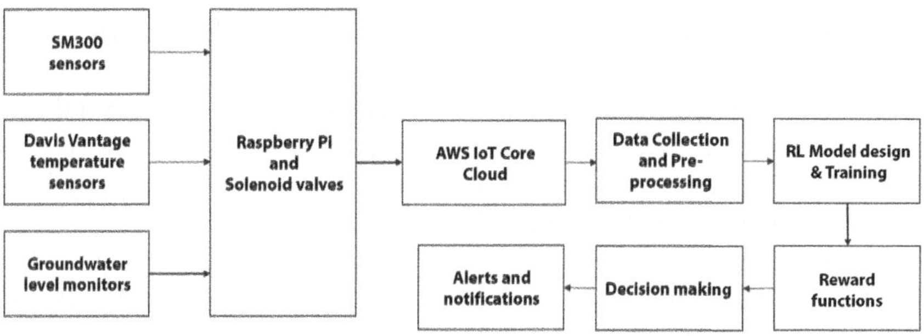


**Figure 10.5** Reinforcement learning-decentralized approach for banana farms.

The system is active with feedback from farmers and uses a user-friendly mobile application localized in Tamil. Registered with the app, this gives farmers the most real-time updates on irrigation schedules, water use, weather forecasts, as well as crop performance, so they can make informed decisions. The app alerts one to recommendations and gets everyone involved in groundwater management. The RL model is fine-tuned based on regular farmer feedback on crop health and its productivity to account for the dynamic environmental changes. All of that happens in a framework that includes robust monitoring and evaluation. Groundwater levels and crop yield are continuously assessed along with key performance indicators (KPIs), such as water use efficiency, against the reference case. On the one hand, data from IoT sensors and remote sensing was analyzed to track irrigation patterns and overall system performance, on the other hand. The pre- and post-intervention analysis compares water usage and crop productivity with tangible evidence of the system's impact. This decentralized approach combines advanced technologies with RL algorithms to promote and support sustainable agricultural practice, improve water conservation, and protect the livelihoods of banana farmers, while establishing benchmarks of innovative groundwater management.

### 10.4.3 Case Study: Reinforcement Learning Approach for Environmental Air Quality Monitoring

An urban RL-based air quality monitoring system is designed to minimize pollution levels efficiently using advanced technologies and machine learning (Figure 10.6). First, IoT-enabled air quality sensors, such as DustTrak and Aeroqual, are deployed, strategically placed at pollution hotspots to measure the concentration of key pollutants, including PM2.5, PM10, $NO_2$, $SO_2$, CO, and $O_3$. These are supported by meteorological sensors, e.g., Vaisala weather transmitters, that measure environmental parameters are temperature, humidity, wind speed, and wind direction. Further, GPS-based or smart cameras are used to gather traffic data to characterize vehicular emission patterns.

These devices transmit the data they collect in real-time to a centralized cloud platform through reliable communication networks like LoRaWAN, 5G, or Wi-Fi. After they are received, the raw data is preprocessed to filter out noise and put it on the same standard for analysis.

An optimization of pollution control strategies is done using an RL model, i.e., deep Q-learning or proximal policy optimization. Historical data is used to learn the RL model and then simulate different scenarios

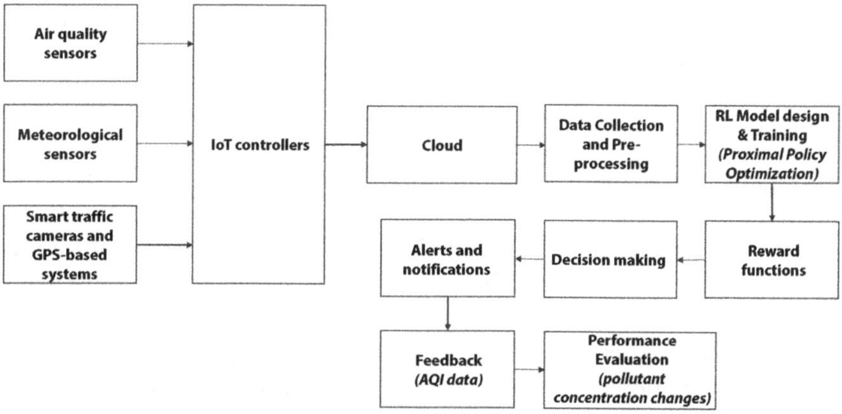


**Figure 10.6** Reinforcement learning approach for environmental air quality monitoring.

to find which actions are most effective. The action space of the model is the regulation of industrial emissions, public alerts, adjusting traffic flow, etc., and the state space includes the air quality parameters, weather conditions, and traffic density. A machine learning-based reward function is built to penalize decisions that worsen air quality and reward decisions that improve air quality. The RL model can be trained, after which it processes real-time sensor data to perform optimal interventions, which are then realized *via* actuators. Based on that, these actuators can act on elements such as traffic signals to reduce congestion, industrial systems for limits on emissions, and public alert mechanisms to alert residents to air quality conditions and desired safety precautions. The system can adapt to new information, as well as a feedback loop. There is real-time AQI data that can continuously check how effective the actions implemented are. The RL model retrains itself when desired outcomes are not met, adjusting based on the feedback to optimize decision-making. The whole system is embedded in a user-friendly interface, like a mobile app or dashboard, enabling them to track air quality, receive alerts, etc. Metrics like reduction in AQI levels, pollutant concentrations, and public compliance rates are analyzed to evaluate the system's performance. This RL-based system presents an integrated strategy for reducing urban air pollution and enhancing health, addressing environmental and public policies while incorporating real-time monitoring, adjustments, and data-driven decision-making.

### 10.4.4 Case Study: Reinforcement Learning-Based Real-Time Flood Management System

An innovative RL-based approach to mitigate flood risks using the Real-Time Flood Management System, which can adapt to real-time data monitoring, historical analysis, and advanced machine learning (Figure 10.7). First, the system gathers critical information input from different sources, such as IoT sensors detecting rainfall intensity, water level, soil saturation, and weather conditions, and a list of past flood patterns and river discharge rate. This is a crucial dataset for understanding the contemporary flood risk landscape. After the data is collected, the system starts by instantiating the operational environment with state variables and determining what actions are possible: opening and closing gates; starting and stopping pumps. A reward function to reward actions minimizing flood risks and structured damage to infrastructure while minimizing response delay is established. After this initiative, environmental conditions are constantly monitored by strategically deploying IoT sensors at places prone to environmental disturbances, such as rivers or urban lowlands. Real-time data from these sensors is transmitted through reliable communication networks to a central server, which preprocesses the information to remove noise, handle missing values, and normalize inputs for analysis. Finally, the preprocessed data was integrated with historical records and weather forecasts to generate a robust dataset on which the RL model can base its decision-making.

The RL model is designed with a clear structure: It represents the current flood environment using real-time sensor data, defines the action

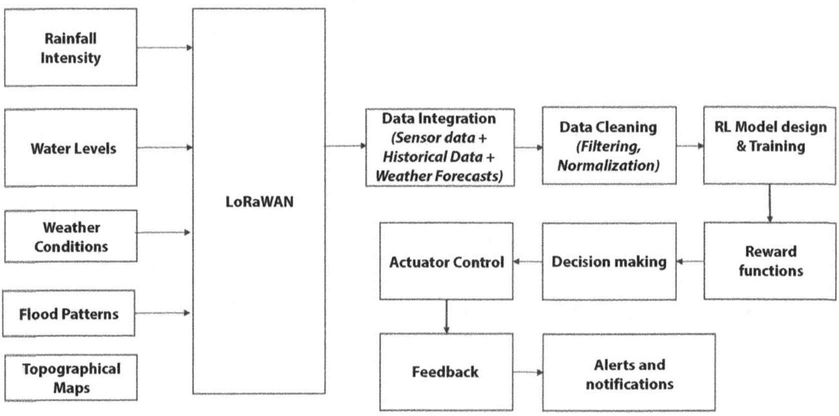


**Figure 10.7** Reinforcement learning-based real-time flood management system.

space in different flood mitigation strategies, and constructs a reward function based on minimizing flood risk and achieving optimal response measures. The model was trained with historical scenarios and simulations to develop efficient control policies that are robust to changing conditions. The RL model is trained once, in which it learns to perform optimal actions for flood mitigation, and the system continuously receives real-time sensor information. The system conducts assessments and executes recommended actions *via* actuators, including dam gates and pumps; should it identify a high flood risk, public alert systems will be turned on to warn residents of potential danger. A feedback loop is integral to the system's operation: It continuously tests the results of actions undertaken on water levels and overall flood conditions, and utilizes this feedback to dynamically refine decision-making capabilities in the RL model. Metrics such as response time and water level reduction are evaluated to assess the performance of the system, and this process of continuous improvement is achieved by lessons learned in each event. With this smooth approach, the flowchart visualizes this entire process, from input collection through decision-making and actuator control to feedback monitoring and evaluation, indicating the role of each module in the system, starting from the input, through the decision-making, entering the actuator control, and finally the feedback monitoring and evaluation sections. This real time flood management system ultimately serves the goal of saving our communities from flooding. However, it also supports building more resilient urban infrastructure that can contend with climate change.

## 10.5 Conclusion

In the context of RL, its application to environmental issues, and more specifically, environmental monitoring, precision farming is another area where a great deal of work is already being done. This work aims to improve the quality of resource management and protect the environment while also enabling industries to become more efficient. The RL-based systems incorporated into user interface design have disruptive capabilities in diverse fields due to the potential offered by data-driven approaches, the IoT-connected sensors, and scalable cloud-edge computing architectures. In the case of agricultural precision farming, irrigation, DQNs frameworks deeply target water resources for irrigation and simultaneously seek to maintain crop yields. These systems optimize and conserve natural resources while ensuring food security. These systems use sensor data, advanced reward functions, and farmer inputs to achieve a dynamically pro-efficient response to environmental changes, which promotes adaptability and efficiency. Likewise, using MARL algorithms in RL environmental monitoring systems for air quality management, the algorithms can achieve limited global and return to local points in the environment without compromising the accuracy and effectiveness of the system. The multi-agent algorithms focus on real-time pollution scenarios and urban congestion levels to achieve better strategic decisions. All these solutions highlight the necessity of machine intelligence to promote resource efficiency, effective systems, and combat climate change. Future work needs to integrate and optimize data while also enhancing computational efficiency, explainability, and ethics. Reinforcement learning technologies can facilitate resource management, environmental protection, and overall quality of life due to their advanced performance level. Solving these challenges is essential for RL systems to change the world positively. And that new world must be developed on sustainable principles.

**AI disclosure statement:** The author confirms that no artificial intelligence (AI)-generated content or AI-assisted writing tools were used in the preparation of this chapter (including summaries and the abstract). All text, analysis, and conclusions are the result of the author's original work.

## References

1. Lobell, D.B., Burke, M.B., Tebaldi, C., Mastrandrea, M.D., Falcon, W.P., Naylor, R.L., Prioritizing climate change adaptation needs for food security in 2030. *Sci.*, 319, 5863, 2008.
2. McBratney, A., Whelan, B., Ancev, T., Bouma, J., Future directions of precision agriculture. *Precis. Agric.*, 6, 7, 2005.
3. Mishra, A.K., Moss, C.B., Erickson, S.W., Technology adoption in precision agriculture and its impact on farm profitability: A review. *Agric. Syst.*, 174, 30, 2019.
4. Gebbers, R. and Adamchuk, V.I., Precision agriculture and food security. *Sci.*, 327, 5967, 2010.
5. Kernecker, T.P., Knierim, J., Wurbs, H., User perspectives on precision farming: A systematic review. *Precis. Agric.*, 21, 1327, 2020.
6. Li, H., Ma, Y., Zhang, J., Ma, X., Decentralized decision-making in smart agriculture: A review. *IEEE.*, 8, 71818, 2020.
7. Villa-Henriksen, F., Edwards, J.E., Pesonen, L.A., Green, D., Sørensen, S.J., Internet of Things in precision agriculture: Applications and challenges. *Biosyst. Eng.*, 191, 60, 2020.

8. Sutton, R., Barto, A.G., Bach, F., *Reinforcement Learning: An Introduction*, MIT Press, Cambridge, MA, USA, 2018.
9. Abdel-Basset, M.A., Manogaran, G., Gamal, M., Chang, T., IoT and AI-enabled smart agriculture: A comprehensive review. *IEEE*, 7, 9053, 2020.
10. Ramesh, M.J., Janakiraman, S.V., Sharma, D.R., RL for irrigation management: A case study. *IEEE Trans. Autom. Sci. Eng.*, 18, 1248, 2021.
11. Gautron, R., Maillard, O.A., Preux, P., Corbeels, M., Sabbadin, R., Reinforcement learning for crop management support: Review, prospects, and challenges. *Comput. Electron. Agric.*, 15, 200, 2022.
12. Zhao, B., Zhao, X., Chen, X., He, Z., A review of RL in agricultural systems: Techniques and applications. *Comput. Electron. Agric.*, 191, 106523, 2021.
13. Lal, R., Managing soils for feeding a global population of 10 billion. *J. Soils Water Conserv.*, 74, 29A, 2019.
14. Kamilaris, S.M. and Prenafeta-Boldu, F.X., Deep learning in agriculture: A survey. *Comput. Electron. Agric.*, 147, 70, 2018.
15. Farooq, M., Umer, M.S., Anjum, M.I., Kim, J., Sustainability of precision agriculture: Current status and future trends. *Sustainability*, 14, 1125, 2022.
16. Elavarasan, D., Crop yield prediction using deep reinforcement learning model for sustainable agrarian applications. *IEEE*, 34, 142, 2020.
17. Schmidhuber, J., Deep learning in neural networks: An overview. *Neural Netw.*, 61, 117, 2015.
18. Tey, Y.S., Brindal, M., Djamhuri, T.L., Adoption of precision agriculture technologies by smallholder farmers. *Agric. Syst.*, 123, 130, 2014.
19. Meyer, D., Kugler, P., Wohner, C., Hilty, L.M., Decentralized and collaborative decision-making in precision agriculture: Challenges and solutions. *Comput. Electron. Agric.*, 184, 1243, 2021.
20. Getahun, S., Kefale, H., Gelaye, Y., Application of precision agriculture technologies for sustainable crop production and environmental sustainability. *Sci. World*, 1, 45, 2024.
21. Wang, X., Zhao, X., Zhang, Y., Applications of reinforcement learning in agricultural automation. *Comput. Electron. Agric.*, 200, 106594, 2023.
22. Basso, B. and Antle, J., Digital agriculture to design sustainable agricultural systems. *Nat. Sustain.*, 3, 254–256, 2020.
23. Shah, M.A., Iqbal, M.W., Ashraf, M.U., Islam, M.A., Imran, M., IoT-based smart farming: Applications, technologies, challenges, and future trends. *IEEE*, 9, 12, 2021.
24. Pierpaoli, E., Carli, G., Pignatti, E., Canavari, M., Drivers of precision agriculture technologies adoption: A literature review. *Proc. Technol.*, 8, 61, 2013.
25. Van Tricht, K., Gobin, A., Gilliams, S., Piccard, I., Crop yield prediction using deep learning techniques. *Comput. Electron. Agric.*, 155, 257, 2018.
26. Yuan, Y., Li, L., Zhang, J., Wang, B., Reinforcement learning-based irrigation scheduling for precision agriculture. *Comput. Electron. Agric.*, 185, 106135, 2021.

27. Wu, Y. and Chen, X., Smart irrigation using reinforcement learning with environmental data. *J. Agric. Food. Chem.*, 68, 14750, 2020.
28. Jones, J., *et al.*, Agricultural system models: State of agricultural systems science. *Agric. Syst.*, 155, 269, 2017.
29. Van Hentenryck, B., *et al.*, Multi-agent reinforcement learning in pest management. *AI Agric.*, 3, 50, 2021.
30. Smith, J. and Patel, R., AI-driven crop monitoring: Enhancing precision agriculture with machine learning and remote sensing. *Artif. Intell. Agric.*, 15, 98, 2023.
31. Jain, A.R., Rathi, H.S., Sharma, S.C., Reinforcement learning for water-efficient irrigation in precision agriculture. *Agric. Water Manag.*, 241, 106304, 2020.
32. Sun, X., Xie, J., Karimi, S.A., Wang, C., Deep Q-learning for optimal water usage in smart agriculture. *IEEE Trans Industr. Inform.*, 17, 3103, 2021.
33. Syed, A.M., Sim, K.P., Ooi, S.L., Implementation of precision irrigation system using reinforcement learning and IoT. *Sustain. Comput. Inform. Syst.*, 24, 100274, 2019.
34. Gupta, M.K., Reinforcement learning for autonomous water irrigation management in agriculture. *Comput. Electron. Agric.*, 146, 17, 2018.
35. Singh, R., Patel, D., Kumar, S., Decentralized AI-driven decision support for autonomous farming using IoT and reinforcement learning. *Comput. Electron. Agric.*, 198, 107103, 2023.

# 11

# Case Study on Crop Knowledge Discovery Based on Reinforcement Learning through Normalized, Homogenized, and Integrated Agricultural Data

**M. Nalini[1]*, Kaarthica Gopi[2], S. Sathya Sai Ram[2] and D. Rajesh Kumar[3]**

[1]*Faculty of Electronics and Instrumentation, Sri Sairam Engineering College, Chennai, India*
[2]*Department of Electronics and Instrumentation, Sri Sairam Engineering College, Chennai, India*
[3]*Symbiosis Institute of Computer Studies and Research (SICSR), Symbiosis International (Deemed University), Pune, India*

***Abstract***

Learning more about crops is needed to improve farming methods, support precision agriculture, and make farming more sustainable. This chapter focuses on using reinforcement learning (RL) to discover advanced crop information by merging and balancing agricultural data. This chapter elaborates on how RL agents work with a farm simulator, repeatedly discovering better farming strategies with the help of rewards. Information on soil type, weather changes, and the plant's developmental stages from various sources is blended to ensure the RL model works with more detailed data. Removing biases and similarities in the data makes the model more useful in a range of farming conditions. Reinforcement leaving frame helps optimize vital elements, such as when to irrigate, how much fertilizer to use, and pest control, by following the feedback given to it. As a result of this advancement, RL aids in making better decisions for agriculture, leading to higher yields, more efficient use of resources, and reduced environmental damage.

***Keywords*:** Integrated agricultural data, data normalization, homogenization, crop knowledge discovery, environmental impact, precision agriculture

**Corresponding author*: nalini.ei@sairam.edu.in; ORCID: https://orcid.org/0000-0001-6457-7990

Anitha Velu, Prasanth Aruchamy, Raghu Ramamoorthy, Rajesh Kumar Dhanaraj and Seifedine Kadry (eds.) Climate Smart Agriculture, (219–240) © 2026 Scrivener Publishing LLC

## 11.1 Insights on Machine Learning Techniques in Agriculture

Agriculture is of utmost significance for global food security; it is directly impacted by social, economic, and environmental factors. The quick development of sensors, drones, and satellite technologies has allowed modern agriculture to generate a great deal of data. Market trends, crops' well-being, soil characteristics, and weather conditions found in these datasets can greatly support farmers in making decisions. Although the lack of uniformity and diversity among these datasets provides obstacles when trying to take advantage of valuable knowledge.

Understanding crops and their features is an important research area that greatly helps with making agricultural decisions at both national and regional levels. Getting details right about weather conditions supports farmers in choosing when to plant and what to plant, leading to better management of farmland and supplies [1]. The interest in using machine learning (ML) comes from its ability to sort through complex data and find patterns that traditional statistics might not notice [2].

Studies today have highlighted that some ML approaches accurately predict crop yields [3]. The review reported an ML technique for judging nitrogen levels in crops, suggesting that further advances in sensor and ML technology may lead to budget-friendly products for farming [4]. Machine learning models are investigated in this chapter to study the impact of climate factors, apart from what traditional models do [5, 6].

Studying how ML is being used in agriculture, such as with crop and animal management. The outcome shows that ML can help solve many agricultural issues. Moreover, fruit ripeness gauging using ML reflects the usefulness of this technology in improving when and how harvests are conducted, which straightaway relates to production prediction [7].

Reinforcement learning allows learning about crops to be improved by handling and integrating agricultural data, supporting flexible and changing ways of learning. While old data mining used unchanging data sets, RL depends on rewards and has an agent responding to a wide variety of circumstances in an agricultural environment. For example, an RL model can learn the best ways to plant by being rewarded for success [8]. Because everything related to soil, weather, and insects is always constant in the data, you can find links in the data that more traditional models might overlook. To manage precision agriculture, it is essential to understand how many factors interact. Moreover, using a variety of different methods, RL supports the creation of creative solutions for farming problems.

## 11.2 Methodologies for Artificial Intelligence-Driven Crop Knowledge Discovery

Improving the quantity of food or crops produced per unit of land is the main issue in agriculture [9]. The necessity of modernizing how we farm and use the latest technologies arises from the sudden and severe changes in the climate. Precision agriculture relies on artificial intelligence (AI) to observe and change different factors that affect crop development. Using regression models, deep learning, and decision support systems (DSS) was one way to support farmers and help them raise their yields quickly. A few techniques are explained.

**Regression Model:** This method starts with cleaning and sorting the data to make sense of it. Linear regression is then used to look at how crop yield might change as important things like weather or rainfall, which are called independent variables, go up or down. The relationship is shown in Equation 11.1.

$$Y = a + (b \times X) + e \tag{11.1}$$

In this, $Y$ denotes crop yield, $X$ denotes something like temperature, $a$ shows the starting point, $b$ is the line's slope, and $e$ stands for the remaining unexplained errors. To see how well the model works, we use measures such as $R^2$ and RMSE. A high $R^2$ value along with a lower RMSE means the model fits the data well and does a good job showing the connection between the different variables.

**Deep Learning:** Gathering a strong dataset comes first in deep learning, as it helps the model make better predictions. The data is cleaned and preprocessed after being drawn together for analysis. Then, the most important factors from the data, for example, include the type of crop, the kind of soil, the temperature, and the amount of rainfall. To perform and simplify better, important features are selected using this method. Convolutional neural networks, recurrent neural networks, and hybrid models are just some of the important deep learning techniques that may be applied here [10]. All these models are validated using the mean absolute error (MAE), mean squared error (MSE), and root mean squared error (RMSE) after the training process. Finally, the model will be used to predict the crop yields expected from the collected data.

**Decision Support Systems:** This system is very useful in agriculture by collecting and organizing data from different farming lands. By using this, farmers can decide with high accuracy. These devices especially help farmers overcome complex issues they encounter in farming. You will often find the following examples of DSSs applied in agriculture:

- Decision Support System for Agrotechnology Transfer
- Soil and Water Assessment Tool
- Farm Advisory Services for Management and Sustainability
- Cropping Systems Simulation Model
- Smart Farming DSS
- GreenSeeker Technology

These systems empower farmers by providing actionable insights to improve effectiveness and sustainability in their agricultural practices.

### 11.2.1 Challenges

Several challenges prevent current agricultural techniques from working well with modern farming. Using fixed datasets that do not keep pace with how quickly agricultural landscapes can change is a main issue. The health of plants, insects on farms, groundwater, and weather conditions can all experience fast changes, but conventional farming usually cannot respond rapidly to these changes. This lack of capability to adjust limits the use and importance of these models in real-world farming situations when fast decisions significantly matter.

A big issue is that ML methods that require labels often lack labeled datasets. Getting labeled agricultural data is often expensive, time-consuming, and challenging in places where there are fewer resources [11]. When there are not enough accurate or complete labeled examples, these methods cannot be as effective or scaled up to meet the needs of various agricultural settings.

In fact, data received from farmers is varied, which causes numerous challenges in merging it. Not all data from multiple sources, such as market data, drones, satellite images, Internet of Things (IoT) devices, and field surveys, is presented in the same way. This type of data is hard for traditional systems to join and analyze, and as such, the analysis may be partial or inaccurate. Most of the time, they focus on achieving quick results and skip the long-term impact on the environment and the preservation of resources. Using current algorithms is often challenging because they require a lot of computing power. Working with large amounts of data in farming tends to require much computing power for analysis. Because of this, using these methods is problematic for farmers, mainly in areas that lack many resources. Furthermore, most of these methods cannot be easily understood and do not provide implementable ideas, making them less valuable for solving real problems in agriculture.

## 11.3 Implementation Reinforcement Learning

### 11.3.1 Agricultural Data Preparation

Before employing data-driven solutions for agriculture, you need to first organize your agricultural data. To create data, raw data must be cleaned, fitted together, converted, and set up in an understandable and arrangeable way. Sensors, space-based satellites, stationary weather outlets, and teams conducting field surveys all contribute different types of agricultural data. Due to this variety in technology, it becomes complicated to ensure that data is easy to share and use. Cleaning is only the first part of preparing data correctly. Other work includes sorting out missing items, correcting errors, ensuring data is similar, standardizing measurements and scales, and taking care of strange or severe results. The industry gathers data from various sources, for example, weather forecasts and soil descriptions, so all the data must be coordinated. Raw data becomes useful in research after being converted into drought or vegetation indices. When data is well-prepared, it is easier to analyze and apply to decisions, helping to achieve higher yields, improved sustainability, and forward-looking farming.

#### *11.3.1.1 Types of Agricultural Data*

Agricultural data involves many areas and forms to show the detailed work of farming and its environment. Here are the main areas that agricultural data cover:

- **Weather Data-** Includes temperature value, precipitation content, humidity, wind speed, solar radiation, and other meteorological variables.
- **Soil Data-** Comprises information about soil properties such as quality, pH value, organic substance content, moisture levels, and nutrient availability.
- **Crop Data-** Includes data on crop types, growth stages, yields, diseases, and nutrient requirements.

- **Remote Sensing Data**- Captured through satellites and drones, this data provides spatial and temporal information on vegetation indices (e.g., NDVI), land cover, and crop health.
- **Sensor Data**- Collected by IoT devices deployed in fields to monitor soil moisture, temperature, pH, and other variables in real-time.
- **Market and Economic Data**- Includes information on crop prices, demand trends, input costs, subsidies, and trade policies.
- **Geospatial Data**- Provides information on topography, land use, and field boundaries.
- **Pest and Disease Data**- Tracks pest populations, disease incidence, and outbreaks.
- **Historical and Policy Data**- Includes past agricultural trends, climate records, and policy frameworks affecting agriculture.
- **Genetic and Genomic Data**- Focuses on crop genetics, seed varieties, and breeding information.

#### *11.3.1.2 Data Cleaning*

All information collected from sensors, weather stations, and satellites in agriculture must be cleaned to ensure the results are precise and dependable. The data is now properly aligned for analysis to assist in decision-making in agriculture. During data cleansing, managers must focus on missing data that is normally caused by sensor failures, environmental changes, or transmitting too little data. Filling these gaps is done with interpolation, imputation, or by excluding a few cases. Duplicates need to be removed as they add unnecessary information and may hide the main points of the study. To ensure accuracy, the procedure also includes mistake correction. Human errors, wrongly filed data, or unusual cases are examples of errors that might arise. Anomalies in data are found by applying statistical methods and understanding the type of data. Reconciling the use of various measurement units in data is necessary by making sure that all datasets contain information in the same type of units.

Finally, fixing discrepancies across datasets brings data gathered from various sources into alignment while considering variations in time zones, geographical locations, and labeling methods. These steps collectively enhance data quality, ensuring robust and reliable insights for agricultural decision-making.

### 11.3.1.3 Data Normalization

The normalization of agricultural data means making different types of data comparable by putting them on a common scale. Since agricultural data often comes from a variety of sources with various ways of measuring temperature, soil moisture, or yield, normalization is needed to compare them all. This ensures that all variables in the data are considered equally, regardless of their range, and allows ML procedures to process the data well. Based on the data and what you want to do, popular normalization approaches include min-max scaling, z-score standardization, and log transformation. The most suitable approach will be based on the details of the agricultural data and the requirements of the model. It is best to use z-score standardization for agricultural datasets involving variability in the data, as it allows for better comparison among the features. It is crucial for ML and predictive analytics that soil properties, temperature, and crop yield are all on the same scale; z-score standardization helps achieve this [12].

Z-score standardization alters the value of data to attain a zero value mean and a standard deviation value of 1. This technique is applicable when the data follows a normal distribution and you need to report variations in scale across various variables. Z-score standardization is depicted in Figure 11.1, which is frequently used in crop knowledge discovery when handling data that has notable variance (e.g., crop yield data, plant growth measurements) and where comparisons across variables with several units are required. It also holds outliers greater than min-max scaling, as it focuses the data around the mean. The formula of z-score tuning is:

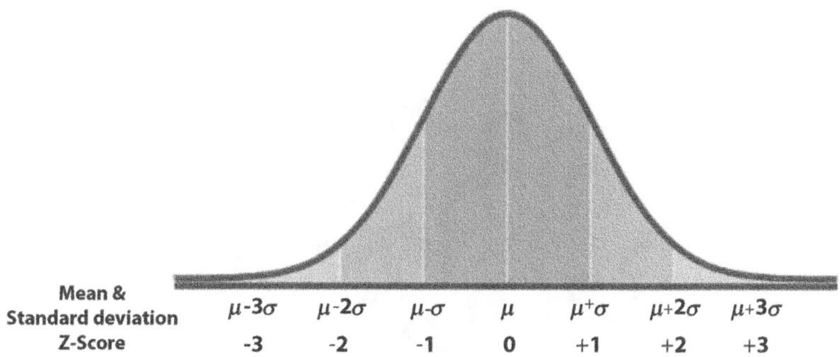


**Figure 11.1** Z-score standardization.

$$Z = \frac{X - \mu}{\sigma} \tag{11.2}$$

Here, $X$ symbolizes observed value, $\boldsymbol{\mu}$ represents the value of the mean, and $\boldsymbol{\sigma}$ represents standard deviation.

#### *11.3.1.4 Data Homogenization*

Agricultural data homogeneity refers to the consistency and uniformity of data across different sources and formats. In agriculture, data often comes from numerous sources, such as sensors, satellite images, weather stations, and field surveys, which can vary in units, scales, and measurement techniques [13]. Achieving homogeneity in agricultural data involves standardizing these varied datasets to ensure they are comparable and compatible for analysis. This includes joining together data taken at different times or areas, creating a regular scale for different measurement units, and ensuring that all the data categories are the same (e.g., making sure bits of numerical data are accurate). When the data is homogenized, the analysis works better, leading to useful decisions about crops, insects, watering, and other farm-based tasks [14].

#### *11.3.1.5 Data Integration*

Using information from weather stations, crop management systems, soil sensors, and satellite images together and in one format for study is called agricultural data integration. As shown in Figure 11.2, the data types are

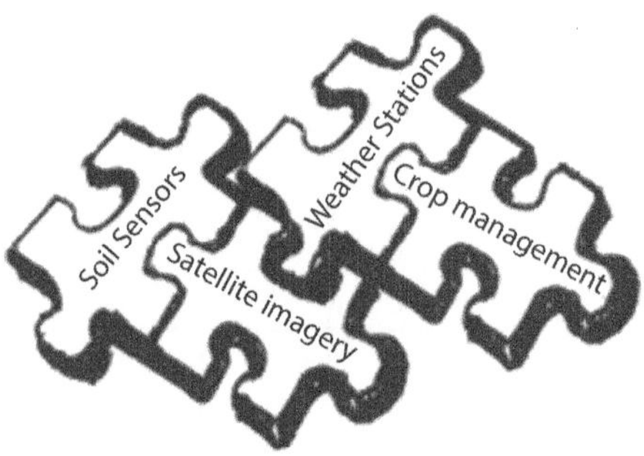


**Figure 11.2** Data integration.

present through combining geographical, temporal, and environmental data; this approach allows for a thorough understanding of different parts of agricultural systems [15, 16]. Aligning how data appears, the sizes of files, and the time frames they cover, plus ensuring compatibility between different databases, is crucial in data integration. By offering a thorough understanding of variables, including soil health, weather patterns, irrigation requirements, and pest management, effective integration enables more precise forecasts and decision-making. By producing a coherent and useful dataset that can be utilized to optimize agricultural techniques and boost production, it strengthens the potential of data-driven strategies, such as ML and DSS [17].

## 11.3.2 Reinforcement Learning

Reinforcement learning is one of the approaches with prominent assurance in AI, especially for applications in decision-making under complex and dynamic environments. In contrast to supervised learning, where models are trained based on labeled data, the agent acquires knowledge through a form of interconnection with its environment by obtaining feedback in terms of rewards or penalties. This ability to consistently learn from experience and adjust to modifying conditions has made RL a flawless candidate for solving many real-world problems, compromising those in the agricultural sector.

The decision in agriculture is almost always complex, coupled with being dependent on weather, soils, pests, and all other growth stages for crops; hence, the application of RL could mean so much in streamlining crop management, especially through improving resource usage efficiency coupled with sustainability. These insights provide a very profound overview of RL, the available types, the utilized methods, and applications. This pertains particularly to crop management.

### *11.3.2.1 Basics of Reinforcement Learning*

Reinforcement learning is a kind of ML technique in which an agent studies to work alongside an environment to maximize progressive rewards through trial and error. Unlike supervised learning, the model is educated on a labeled dataset; RL agents gain knowledge by exploring and taking actions in an environment, with feedback from the outcome of those actions.

#### 11.3.2.1.1 Key Components of Reinforcement Learning

1. **Agent**: The party who communicates, leading to some decisions or actions to attain a goal. Here, the agent may act as an AI system suggesting irrigation timings or controlling pests.
2. **Environment**: How the environment around the agent affects its behavior. This can be either a real farm or a simulation, where an agent collects data and uses it to make good decisions.
3. **State *(s)***: It reflects the environment's current state. As an illustration, during crop management, the state can consider soil moisture, temperature, crop health, and weather.
4. **Action *(a)***: Moves and choices made by the actor. Examples of actions in agriculture are watering crops, adding fertilizers, or carrying out pest control.
5. **Reward *(r)***: Reward is the outcome you get after choosing a certain state. Some factors that make up rewards in crop management are crop yield, use of water, pests, and use of energy efficiently.
6. **Policy *(π)***: The strategy or plan used by the agent is to decide which action to take in any given state. The policy finalizes the best action based on the current state.
7. **Value Function *(V(s))***: Based on state '*s*', the value function measures how high the cumulative reward can be for the agent. It makes it easier to decide which states the agent should target.

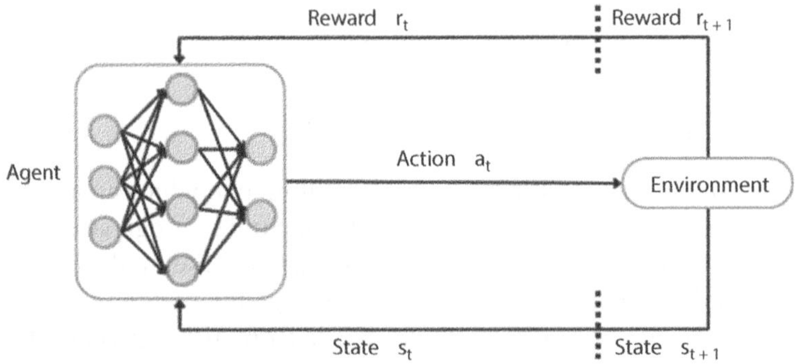


**Figure 11.3** Reinforcement learning.

8. **Q-Function** ***(Q(s, a))***: This Q function calculates the expected cumulative reward for a given action 'a' taken in state 's'. The agent uses this function to decide which actions to take in any given state.

Figure 11.3 shows the pictorial representation of RL. Agents in RL aim to learn an optimal policy that improves the total cumulative reward over time. The learning process involves the following steps:

**Exploration:** This means that the agent randomly discovers the environment to gather information about different states and actions.
**Exploitation:** The agent starts leveraging the knowledge it has acquired by selecting the optimal actions based on the current state.
**Reward Feedback:** After performing some action, the agent collects a reward (or penalty), which informs whether that action is necessary or not.
**Learning and Optimization:** The agent uses the reward feedback to improve its learning in that environment and enhance its policy over time.

#### 11.3.2.1.2 Types of Reinforcement Learning

**Model-Free RL:** Experience leads the agent to learn facts straight from reality, without first constructing a model. Instead of trying to guess how the system will move forward and what it will get as rewards, this kind of RL works on choosing the best action by testing out different options. Q-learning and SARSA are considered two popular model-free algorithms.
**Model-Based RL:** The agent learns things by doing them rather than trying to make a model of its surroundings. This type of RL means the agent tries to learn the rules and what happens to the environment as it changes, as well as what rewards it can get. With this model, the agent can look ahead and decide what it should do next. This approach is harder to compute, but it can work faster when there's a model or something similar that one can learn or use instead of trying to figure out the answer.
**Value-Based Methods:** Valuation-driven approaches allow the agent to learn the value of states and their actions. An often-used algorithm in this category is Q-learning, which helps the agent learn an approximate value for what it will get as a reward for each action taken in every state.
**Policy-Based Methods:** They are concerned with finding a direct way to translate each state to the right action, which is a policy. The agent is taught to seek out a higher reward by optimizing with methods such as gradient ascent. An example of such an algorithm is Reinforce, which can optimize policy directly instead of creating an estimate for the value function.

**Actor-Critic Methods:** Actor-Critic methods are a mix between value learning and policy learning. It is up to the actor to choose which action to take, and the critic observes those actions and measures them with a value function. Often, these methods are more reliable and effective than others, as shown by algorithms like A2C Actor-Critic.

### 11.3.3 Deep Q-Learning for Crop Knowledge Discovery

Using RL improves decision-making in complicated and confident environments, especially as it relates to precision agriculture. Deep Q-learning is an RL technique that pairs Q-learning with the strengths of deep neural networks. It allows the discovery of the best ways to manage crops by looking at old patterns and up-to-date conditions. Deep Q-learning, when used in agriculture, may strengthen irrigation, add fertilizer, and control pests.

#### *11.3.3.1 Fundamentals of Deep Q-Learning*

##### 11.3.3.1.1 Q-Learning Overview

Q-learning is a model-free RL procedure in which an agent learns to take actions that amplify a cumulative reward over time. In traditional Q-learning, the agent uses a Q-function. It evaluates the expected reward for a given state-action pair. This Q-learning is used to find the best policy that increases the final reward. The Q-value update rule is as follows:

$$Q(s_t, a_t) \leftarrow Q(s_t, a_t) + \alpha \left(r_t + \gamma \; max \; Q(s_{t+1}, a) - Q(s_t, a_t)\right) \tag{11.3}$$

Where:

- $Q(s_t, a_t)$ is the Q-value for taking action in the state.
- $r_t$ is the reward.
- $\gamma$ is the reduction factor used for future rewards.
- $\alpha$ is the learning rate used to control the Q value.

In the traditional Q-learning method, this process required maintaining a Q-table for every possible state-action pair, which is impossible for environments with large or continuous state spaces, such as agriculture.

#### 11.3.3.1.2 Deep Q-Learning: Extending Q-Learning with Deep Neural Networks

Deep Q-learning overcomes the limitations of traditional Q-learning by resembling the Q-function using deep neural networks. Instead of keeping a large Q-table, the Q-values are characterized by a neural network that maps states to action values. This allows DQN to become high-dimensional, continuous state spaces, making it suitable for real-world applications, such as crop management [18].

The Q-function estimation in DQN is:

$$Q\ (s,\ a;\ \theta) \approx Neural\ Network\ (s,\ a;\ \theta) \tag{11.4}$$

Where:

- $\theta$ weight of the neural network.
- $Q\ (s,\ a;\ \theta)$ predicted value of Q

**Neural Network Architecture**

The DQN neural network typically involves several layers of neurons shown in Figure 11.4:

- **Input Layer**: State information is received by input layer, for example in agriculture data like soil moisture levels, weather situations, crop health, and other environmental factors.
- **Hidden Layers**: The network processes this input through one or more hidden layers, which allow the agent to acquire complex, non-linear relationships in the data.

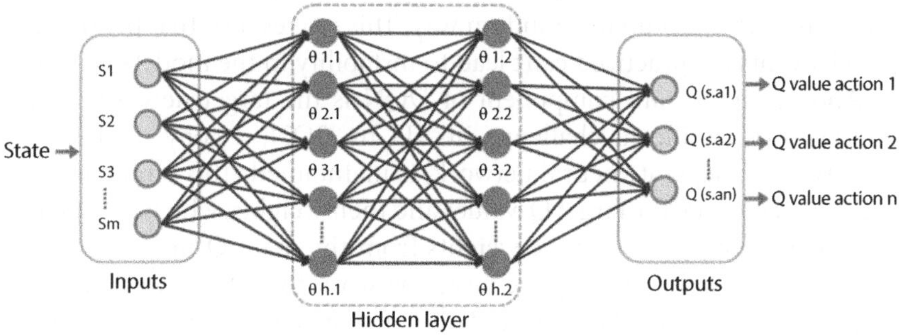


**Figure 11.4** Deep Q-learning.

- **Output Layer**: The output layer consists of one neuron for each possible action the agent can take, where the output corresponds to the Q-value for each action.

#### 11.3.3.1.3 Deep Q-Learning Algorithm

Deep Q-Learning, an RL algorithm, accumulates the power of the Q-learning primary technique used in RL with the dominant responsibility approximation capabilities of deep neural networks. That is, it will undertake intricate environments with high-dimensional state spaces, such as video games, robotic control tasks, ahead of the scope of tabular traditional Q-learning. The predominant idea is to make use of the deep neural network to estimate the Q-function, which evaluates the anticipated collective reward for taking a specific action in a given state and then follows the optimal policy thereafter. This approach makes the agent generalize over similar states; even if it must handle vast or frequent state spaces, it can acquire knowledge about policy.

The first step in DQN is to initialize the Q-function approximation as a deep neural network with random weights, denoted by '$\theta$'. The network uses current '*s*' (state) as input and provides Q-values for any movement in that particular state. The goal is to keep improving '$\theta$' until the network's predictions of Q-values match the true values calculated with the equation. It uses this network to help choose actions that will give the most rewards in the future.

Experience replay is an important improvement in DQN, as it helps the learning process to be both stronger and more productive. As agents go through training, they engage the environment and collect what they learn in a replay buffer. Whenever an experience is created, it describes the existing state '*s*', the action '*a*', the reward '*r*', and the resulting state '*s*.' Instead of using the recent experiences for learning, the agent chooses some experiences from the backup in a random way. This makes it so that the order in which events are practiced can't lead to instability in the model's learning.

Another important component of DQN is the target network, which addresses a significant challenge in RL. In DQN, a copy of the main Q-network, a separate neural network, is maintained. This target network is used to compute the target Q-values and their weights, which are updated less frequently by copying the weights from the main Q-network at consistent intervals. By holding the target network fixed for several updates, DQN ensures that the target values do not change too rapidly, providing a more stable optimization process. This separation of roles is a main feature of DQN.

The agent selects actions using an epsilon-greedy policy, a strategy that balances new actions (exploration) and choice of effective action (exploitation). At each step, the agent determines a random action with probability '$\epsilon$' to explore the environment and potentially discover better strategies. With the probability "$1-\epsilon$", the agent chooses the action with the peak predicted Q-value, manipulating its current knowledge of the environment. Over time, the value of $\epsilon$ decays, shifting the focus from exploration to exploitation as the agent studies more about the environment.

The central learning process in DQN includes updating the Q-function based on the Bellman equation, which shows the relationship between the Q-values of successive states. Specifically, the algorithm minimizes a loss function that measures the transformation between the predicted Q-value $Q_\theta(s, a)$ and the target Q-value.

Loss function can be minimized by regulating the weights $\theta$ of the main Q-network using stochastic gradient descent or a similar optimization algorithm. The network learns to approximate the value of the optimal Q-function by iteratively reducing the error between the predicted and target Q-values.

Training progresses over many episodes, where each episode involves the agent interacting with the environment from an initial state until it reaches a terminal state or a maximum number of steps. The agent observes the state, takes actions based on its epsilon-greedy policy, receives rewards, and transitions to new states. At each step, it stores experiences in the replay buffer and periodically updates the Q-network using mini-batches of experiences. The target network is reorganized less frequently to ensure stability.

Several improvements have been anticipated to further improve the performance of DQN. For example, double DQN mitigates the overestimation bias in standard DQN by decoupling the action selection and evaluation steps. In double DQN, the main Q-network selects the action with the highest Q-value, but the target network evaluates its value. Another enhancement, ordered experience replay, allocates higher probabilities to experiences with larger temporal-difference errors, ensuring the agent focuses more on experiences that are difficult to learn.

Deep Q-learning has been proficiently applied to a wide range of challenging tasks. For instance, it attained human-level performance on multiple Atari video games, a benchmark for RL algorithms. Beyond games, DQN has been used in robotics, where agents gain knowledge to govern robotic arms or navigate environments, and in real-world decision-making problems, such as optimizing traffic signals, resource allocation, and even financial portfolio management.

For each time step, the agent updates its Q-values using the Bellman equation, which in DQN becomes:

$$L(\theta) = E[(r_t + \gamma \, max \, Q'(s_{t+1}, a; \theta^{-}) - Q(s_t, a_t; \theta))^2] \quad (11.5)$$

Where:

- $Q'(s_{t+1}, a)$ is the target Q-value, computed by the target network $Q'(s_{t+1}, a; \theta^{-})$.
- $\theta^{-}$ are the weights of the target network.

The objective is to minimize the loss function $L(\theta)$, which is the change between the predicted Q-values and the target Q-values.

#### 11.3.3.1.4 Application of Deep Q-Learning for Crop Knowledge Discovery

In crop management, the goal of using DQN is to optimize various agricultural practices such as irrigation, fertilization, pest control, and crop health monitoring. In this context, the environment is the crop field, and the agent's actions involve deciding on optimal farming practices based on current environmental conditions.

##### 11.3.3.1.4.1 State Representation

The state 's' in the crop management problem has the following features

- **Soil moisture**: The amount of water in the soil directly affects irrigation decisions.
- **Temperature**: The ambient temperature that affects crop health, growth, and water evaporation.
- **Crop health**: Information from sensors or satellite images to determine the current health of the crop.
- **Weather forecast**: Weather conditions like temperature, rainfall, and humidity predictions can influence farming practices.

These states are often represented as vectors or matrices, and the neural network receives this data as input.

#### 11.3.3.1.4.2 ACTIONS

The action *'a'* represents the decisions made by the agent. In crop management, these could include:

- **Irrigation**: The amount of water applied to the crops.
- **Fertilization**: Adjust the amount of fertilizer used.
- **Pest Control**: Requirement of pesticide or other pest control methods.
- **Harvesting**: Suitable time to harvest crops for maximum yield.

#### 11.3.3.1.4.3 REWARDS

The reward *'r'* provides feedback to the agent based on its actions. For example:

- A positive reward could be given for actions that result in increased crop yield or efficient water usage.
- A negative reward could be assigned for excessive water usage, pest damage, or insufficient crop yield.

The reward function can be tailored to encourage sustainability and efficiency in farming practices. The pictorial representation of the DQN application for crop knowledge discovery is shown in Figure 11.5.

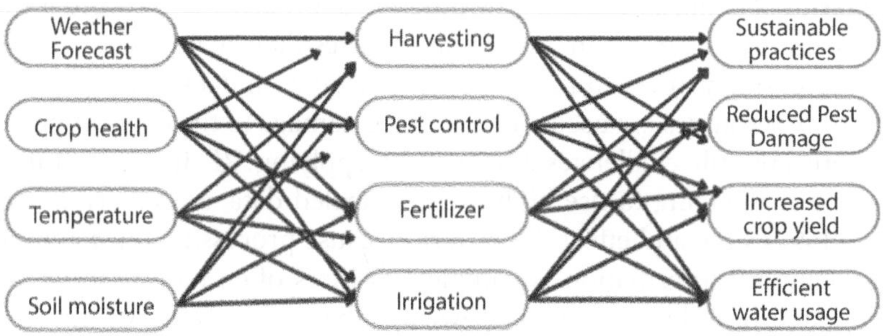


**Figure 11.5** Application of DQN for crop knowledge discovery.

## 11.4 Advancing Agricultural Decision-Making with Reinforcement Learning

Combining RL with the study of crop knowledge is an important step for agriculture that helps with saving resources and improving sustainability. Many research findings suggest that RL can help cope with the challenges in modern agriculture that cannot be handled properly by traditional methods because they rely on unchanging sets of data and cannot handle real-time changes in the environment.

Reinforcement learning is especially useful because it learns effectively as it goes through different interactions on the farm. Because its policies are constantly updated with real-time information, the RL model adapts to changes in the weather, soil moisture, and infestations, helping provide suitable recommendations for crops. Because conditions in farming can rapidly shift, it is necessary to react quickly and make the best decisions to yield the highest results and minimize losses.

Using the same data format for all the agricultural data better trains the RL model. When data is free of errors and biases, the model can work in a greater number of agricultural areas. This stands out because agricultural data usually comes from various sources, including satellite, IoT, and check-in-the-field methods. Combining various data sources provides the model with more accurate insights and also gives a better understanding of the entire agricultural field.

Additionally, RL is used to improve important farming functions such as setting up irrigation, using fertilizers appropriately, and dealing with pests. As a result of being able to imagine different approaches and consider their outcomes, farms can act in a way that matches their main goals. It means that increasingly, precision agriculture is relying on data to increase how efficiently resources are used and to reduce damage to the environment.

There are still challenges involved in applying RL in agriculture. It remains challenging to use high-quality labeled data, as gathering it can take a lot of time and effort. Moreover, because processing huge amounts of agricultural information is complex, farmers often require impressive infrastructure and professional knowledge that is not always within reach. To make RL technologies useful in widely adopting them in agriculture, we first need to deal with these issues.

This study underscores the evolutionary ability of RL in crop knowledge discovery. By leveraging integrated agricultural data and adaptive learning mechanisms, RL can remarkably upgrade agricultural decision-making and guide to enhanced efficiency and sustainability. Future research should pivot on developing user-friendly tools and platforms that clear the way for the adoption of RL in diverse agricultural settings, promising that farmers can harness the maximum benefits of this innovation.

## 11.5 Conclusion

The remarkable ability of RL to help farmers learn more about crops and make farming work better has already been shown. By using organized and similar farm data, RL makes it easier to talk about the challenges and problems that come with today's farming. The potential of RL agents to learn from what's happening in real farms allows them to give useful advice in a short time for important things like water, nutrients, and keeping pests away. The importance of using good quality data and making sure it all fits together properly was made sure to help the RL models work better. By making sure that farming data is collected the same way every time and is free from any biases, the models can work well in different places, which helps people make better choices about farming decisions. This capability is especially important in precision agriculture, since figuring out how all the different parts work together helps farmers get better crops and do less harm to the environment.

Several challenges were taken care of so that using RL in agriculture could work well. The need for really good labeled datasets and the large amount of computing needed to work with a lot of agricultural data make it tough for more people to use these tools. Future efforts should focus on creating easy-to-use tools and systems that help make it simpler for farmers to use RL technologies in their daily work. In conclusion, using RL to learn more about crops shows how farming is adapting with new ways of thinking and technology. By using the power of strong data analysis and learning programs that can adjust to new information, RL helps farmers make better choices, which leads to using resources more efficiently, doing more with less, and helping keep farming sustainable. Continued research and development in this area will be needed so we can clearly see what RL can offer and help farmers deal with the growing complexity in their jobs.

**AI disclosure statement:** The author confirms that no artificial intelligence (AI)-generated content or AI-assisted writing tools were used in the preparation of this chapter (including summaries and the abstract). All text, analysis, and conclusions are the result of the author's original work.

## References


1. Van Klompenburg, T., Kassahun, A., Catal, C., Crop yield prediction using machine learning: A systematic literature review. *Comput. Electron. Agric.*, 42, 277, 2020.
2. Vuong, M.N. and Kechadi, M.T., Crop knowledge discovery based on agricultural big data integration, in: *Proceedings of the 4th International Conference on Machine Learning and Soft Computing (ICMLSC '20)*, pp. 46–50, Association for Computing Machinery, New York, NY, USA, 2020.
3. Al-Adhaileh, M.H. and Aldhyani, T.H.H., Artificial intelligence framework for modeling and predicting crop yield to enhance food security in Saudi Arabia. *PeerJ Comput. Sci.*, 8, 24, 2022.
4. Brown, A.B., Advances in machine learning. *J. Artif. Intell.*, 10, 123, 2022.
5. Chlingaryan, A., Sukkarieh, S., Whelan, B., Machine learning approaches for crop yield prediction and nitrogen status estimation in precision agriculture: A review. *Comput. Electron. Agric.*, 151, 61, 2018.
6. Elavarasan, D., Vincent, D.R., Sharma, V., Zomaya, A.Y., Srinivasan, K., Forecasting yield by integrating agrarian factors and machine learning models: A survey. *Comput. Electron. Agric.*, 155, 257, 2018.
7. Li, B., Lecourt, J., Bishop, G., Advances in non-destructive early assessment of fruit ripeness towards defining optimal time of harvest and yield prediction - a review. *Plants (Basel)*, 7, 3, 2018.
8. Barto, A.G. and Dietterich, T.G., Reinforcement learning and its relationship to supervised learning. *J. Artif. Intell.*, 72, 81, 2024.
9. Sangeetha, R. and Aruchamy, P., Cognitive Intelligence and Distributed Computing Systems Applications in Smart Farming, in: *Computer Vision in Smart Agriculture and Crop Management*, pp. 159–186, Scrivener Publishing LLC and John Wiley & Sons, Inc., 2025.
10. Huang, Y., Sun, S., Duan, X., Chen, Z., A study on deep neural networks framework, in: *IEEE International Conference on Machine Learning*, pp. 145–152, IEEE, Shijiazhuang, China, 52, 2016.
11. Abinaya, M. and Vadivu, G., Role of Big Data Analytics in Intelligent Agriculture, in: *Computational Intelligence in Internet of Agricultural Things*, pp. 19–48, Springer Nature Switzerland, 2024.
12. Al-Faiz, M.Z., Ibrahim, A.A., Hadi, S.M., The effect of Z-Score standardization (normalization) on binary input due to the speed of learning in back-propagation neural network. *Iraqi J. Inf. Commun. Technol.*, 1, 41, 2019.
13. Ujjainia, S., Gautam, P., Veenadhari, S., Crop yield prediction using regression model. *Int. J. Innov. Technol. Explor. Eng.*, 9, 269, 2020.
14. Sathya, S.P., Mohammad Shaheer, A., Dharaneesh, S., RamPrakash, S.U., Season and production-based crop recommendations system for agricultural data using machine learning technique, in: *Proceedings of the KILBY 100 7th International Conference on Computing Sciences 2023 (ICCS 2023)*, pp. 125–131, 2023.

15. Reddy, D.J., Crop yield prediction using machine learning algorithm, in: *Proceedings of the Fifth International Conference on Intelligent Computing and Control Systems (ICICCS)*, pp. 142–147, Anantapur, India, 2021.
16. Cravero, A., Pardo, S., Galeas, P., López Fenner, J., Caniupán, M., Data type and data sources for agricultural big data and machine learning. *J. Sustain.*, 14, 16131, 2022.
17. Smith, J.A. Understanding data sampling techniques. *Data*, 15, 45, 2021.
18. Fragkiadaki, K., Deep reinforcement learning and control, CMU 10-403, 1, 10, 2015.

# 12

# Case Study on Soil Health Surveillance: Establishing Reinforcement Learning for Decision-Making and Improving Product Quality

**M. Nalini[1]*, Kaarthica Gopi[1], Sathya Sai Ram[1] and Mariya Ouaissa[2]**

[1]*Department of Electronics and Instrumentation Engineering, Sri Sai Ram Engineering College, Chennai, India*
[2]*Networks and Cybersecurity, Cadi Ayyad University, Marrakech, Morocco*

***Abstract***
Soil health, which directly affects yield and harvest quality, creates extensive and complex data records with the help of advanced sensor networks and remote sensing. This case study examines the application of increased reinforcement learning (RL) in monitoring soil health to improve the decision-making process and improve agricultural quality. This chapter describes the RL algorithms used to analyze the data and create dynamic models that predict soil attributes such as nutrient levels, moisture, and pH values. The RL model optimizes soil management practices such as fertilization, irrigation, and plant rotation through iterative learning and reward-based feedback. The integration of standardized multisource data enhances model reliability, allowing for accurate real-time adjustments to soil treatment and actively responding to changes in ground conditions, thereby preventing potential losses. The use of RL is not only focused on soil health, but also involved in improving harvest quality and sustainability. This study highlights the potential for RLs to revolutionize soil management, leading to increased agricultural return, efficient resource consumption, and improved product quality.

***Keywords*:** Soil health monitoring, agricultural product quality, sensor networks, remote sensing, soil management practices

**Corresponding author*: Nalini.EI@sairam.edu.in

Anitha Velu, Prasanth Aruchamy, Raghu Ramamoorthy, Rajesh Kumar Dhanaraj and Seifedine Kadry (eds.) Climate Smart Agriculture, (241–260) © 2026 Scrivener Publishing LLC

## 12.1 Soil Health Surveillance

Soil health is a critical determinant of agrarian productivity and environmental sustainability. It directly affects the land's capacity to support crop growth, maintain biodiversity, and contribute to ecological stability. Still, soil systems are innately complex, influenced by a multitude of physical, chemical, and natural factors that interact strongly over time [1–4]. As global challenges like climate change, population growth, and resource failure consolidate, the need for effective soil health operations has become more pressing. Traditional styles of soil health assessment are calculated on homemade slices and laboratory analysis, which can be labor-intensive, time-consuming, and limited in scale. The rapid advancement of digital technologies offers an occasion to revise this field by enabling more precise, effective, and scalable surveillance systems.

Though traditional soil testing techniques remain important, they are often quite costly and time-consuming, thus encouraging a search for more efficient alternatives. Recent studies have particularly emphasized the powerful capabilities of machine learning algorithms applied to complex datasets that account for a wide range of soil properties and environmental factors. In previous research, other models have the capabilities of deep learning, such as convolutional neural networks (CNNs) and long short-term memory (LSTM), in effectively predicting attributes of soils such as organic matter content and nutrient availability. The effectiveness of artificial neural networks (ANNs) was seen in predicting compressibility parameters of soils, while modelling soil water contents using artificial intelligence (AI) techniques. Despite these advancements, a lack of comprehensive frameworks remains, which integrate AI into soil health management practices, particularly in diverse agricultural settings [5, 6].

Artificial intelligence-emphasizing reinforcement learning (RL) is very useful in that paradigm. Reinforcement learning is that subset of machine learning that develops intelligent agents, through an iterative process of trial and error with the environment, that can achieve optimal decision-making ability. So, through RL, soil health surveillance systems can be upgraded from static assessment modes to dynamic decision-making frameworks. These systems will not just monitor soil conditions but, with real-time data and feedback, will also recommend adaptive management strategies. This integration will enhance the efficiency of resource use, improve environmental performance, and boost agricultural productivity.

Also, the connection between healthy soil and the quality of agricultural products is coming into sharper focus. Healthy soils supply more nutrients, hold more moisture, and are less prone to pests and diseases, all key determinants of crop quality. With proper data analysis, soil health indicators can be linked more precisely with the produce end of the supply chain for precision and market-oriented production. This study investigates the use of RL for soil health monitoring and its implications for improved decision-making, aiming to maximize produce quality. This would allow for provision for technical bottlenecks, real-world data analysis, and the generation of alternative methods toward a more sustainable and resilient agricultural future.

## 12.2 Methods Used in Soil Health Surveillance

Traditional styles of soil health surveillance encompass physical, chemical, and natural analyses that have been foundational in understanding soil parcels. Physical analysis includes soil texture, structure, and contraction using instruments like sieves and penetrometers, as well as field tests to measure corrosion and water infiltration rates. Chemical analysis generally includes laboratory tests to determine pH, nutrient levels, and levels of nitrogen, phosphorus, potassium, salinity, and organic carbon content, which are mostly determined *via* wet chemical methods such as titration to assess soil fertility. Biological analysis includes estimating microbial exertion through respiration tests or enzyme assays and assessing biodiversity through earthworm counts and root-associated fungi studies. While these styles offer perfection, their time-consuming and labor-intensive nature, as well as spatial limitations, challenge their scalability for large-scale or real-time operations [7].

Detector-grounded styles offer nonstop, *in-situ* monitoring of soil parcels, surpassing traditional ways in speed and content. Electromagnetic detectors measure conductivity to infer texture, salinity, and humidity. Soil humidity detectors, like capacitance and threat detection and response examinations, give real-time data on water scarcity, while spectroscopy analyzes soil nutrients and organic matter. Remote sensing *via* satellite and drone imagery uses indicators like the normalized difference vegetation index to assess soil parcels, though estimation can vary with soil type and environmental conditions [8].

The Internet of Things (IoT) in smart agriculture enables precise monitoring of soil, rainfall, and crop conditions. Detectors measure parameters, such as soil temperature, humidity, nitrogen, phosphorus, potassium, solar radiation, and soil oxygen, transmitting data to a central system for analysis and decision-making. Soil temperature and humidity are tracked using examinations, affecting root growth, photosynthesis, and nutrient transport. Weather stations collect data on rainfall, moisture, and wind, while nitrogen, phosphorus, and potassium detectors give insight into nutrient situations. The IoT also facilitates robotization, similar to controlling irrigation based on soil humidity, perfecting effectiveness and resource operations.

The methods that use spatial interpolation styles to prognosticate soil parcels across a field, grounded on a set of point samples, are geostatistical [9]. Geostatistics is an important tool in soil science, furnishing insight into the spatial variability of soil parcels. It involves the use of statistical styles to dissect and prognosticate the distribution of soil attributes across different locations. Some crucial styles are:

- Semivariogram Analysis
- Kriging
- Co-Kriging
- Inverse Distance Weighting
- Spatial Retrogression
- Indicator Kriging

The decision support system (DSS) for soil and water conservation agrarian climaxes was developed [10]. It uses Visual Basic to induce decision scripts for managing natural coffers. The DSS considers mean pitch, deposition loss, soil type, and land capability class to suggest structural and cropping measures to reduce soil loss and conserve water. The DSS proposes measures like bench sundecks, graded figure bunds, and conservation dikes, which led to a reduction in soil loss.

A comprehensive review identified the contribution of various machine learning applications in modern agriculture, highlighting their effectiveness in optimizing agricultural practices. The processes stressed ranged from data collection through sensors and training models on selected agricultural datasets, defining features through principal component analysis or independent component analysis, and realizing predictive analytics to improve all facets of crop management, yield prediction, and resource use efficiency.

Various machine learning techniques are being used in agriculture and environmental monitoring to enhance soil quality assessment and management. Deep learning has been applied to complex soil data sets to predict soil quality parameters with high accuracy, aiding sustainable agricultural practices by helping in resource decision-making. The application of ANNs has been important in modelling soil water content at field capacity and permanent wilting points, which can guide efficient water management in different soil contexts [11, 12]. Together with sensor data, machine learning approaches have been widely used in crop yield prediction, improving resource allocation, and, generally, environmental monitoring [13, 14].

Together, these methodologies underscore the role of advanced machine learning models in fostering sustainable practices, efficient resource utilization, and improved crop management through soil health optimization [15].

### 12.2.1 Challenges

Many challenges are involved in soil health surveillance methods, which facilitate their application but not complete efficiency [16]. Commonly used methods of assessing soil health, including physical, chemical, and biological analyses, are tedious, time-consuming, and not practical for proper monitoring of larger areas or the soil health dynamics. Those techniques only offered static snapshots of soil conditions, necessitating resampling every other time, which came with time delays that hindered timely decision-making.

Sensor-based techniques, while offering real-time data, face challenges with sensor accuracy and calibration. Soil properties can vary significantly, leading to inconsistent readings. The high cost of advanced sensors, combined with the need for technical expertise, makes these methods inaccessible for many smallholder farmers.

Data-driven approaches, such as machine learning, demand huge amounts of high-quality data and polished computational resources, which are frequently lacking in multiple regions. Moreover, these models can be strenuous for non-experts to interpret, resulting in hesitation in adoption. A decision support system (DSS) mostly requires specialized training, and resistance to alteration can further delay adoption. Finally, the lack of standardization and interoperability among different techniques complicates integration and data interpretation, making it harder to develop scalable solutions. These challenges highlight the need for more cost-effective, user-friendly, and integrated solutions that bridge the gap between advanced technologies and practical farming needs.

## 12.3 Establishing Reinforcement Learning in Soil Health Monitoring

### 12.3.1 Soil Health Indicators

Soil health is a multivariate and thus complex concept constituted by physical, chemical, and biological characteristics, allowing the soil to work properly throughout an ecosystem. This cannot be derived from a single variable, say, crop yield and water quality. It should incorporate a variety of indicators for assessment. Those indicators, as outlined in Figure 12.1, embody the physical, chemical, and biological characteristics, processes, or properties that can manifest measurable insights on soil functionality. They may also include the visual or morphological traits of plants. A wide understanding of those indicators is of great importance in managing soil health and supporting sustainable agriculture, as it guarantees the long-term productivity of soil and environmental sustainability and balance.

#### *12.3.1.1 Physical Indicators*

Soil physical indicators give precious information about soil health and capacity for plant growth. Bulk density is one among soil compaction and

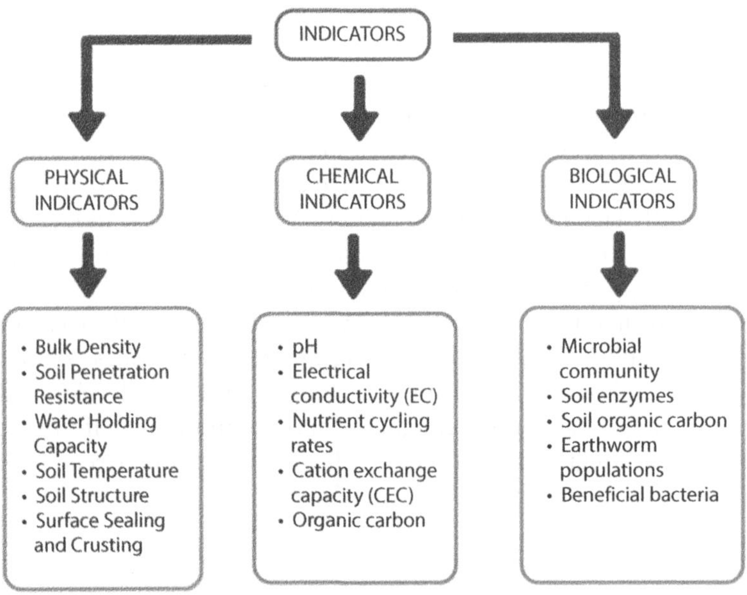


**Figure 12.1** Soil indicators.

porosity, which has direct effects on water, nutrient transfer, and a direct effect on aeration. A lower bulk density is reflective of healthier soil, and often soil with a reduced bulk density is used under conservation tillage by incorporating crop residues. Soil penetration resistance is a measure of the ease of penetration of the root configurations, whereas high resistance means compaction, which can cause problems with both water infiltration and root growth. The retention of moisture within soils is critical to the healthy growth of any class of plants; with biological activity, improved structure, and texture are normally exhibited in soils of either organic matter-rich origins or added in biochar. Soil temperature is the major moving factor encompassing physical, chemical, and biological processes of concern for the growth of plants and the activities of microorganisms. Variations in properties of soil temperature properties arose from solar radiation, moisture variations, and seasonal fluctuations. Soil structure refers mainly to the arrangement of soil particles into aggregates, which becomes basic for water infiltration, root development, and also for microbial activity; this has been improved mainly through conservation agriculture practices such as crop rotation. Surface sealing and crusting, on the other side, is a result of aggregate breakdown, thus affecting infiltration rates, inciting runoff and erosion; hence, it could be enhanced through sustainable soil management practices.

#### *12.3.1.2 Chemical Indicators*

Chemical indicators provide a means for assessing the health of the soil, affecting the availability of nutrients, plant growth, and biological processes in soil. The pH, considered a master soil variable, has a great influence on microbial activity and enzyme efficiency, in addition to the availability of all three nutrient elements-nitrogen, phosphorus, and potassium-added to the soil. When optimal, the pH maintains balanced soil chemistry; extremes that can be toxic to plants signify disruption in the balance. Electrical conductivity measures salinity in the soil based on the ability to conduct an electrical current. High electrical conductivity reflects salinity problems that could also stress out plants by blocking water uptake, while low conductivity means low nutrient availability. The nutrient cycling rate shows how fast nutrients like nitrogen, phosphorus, and potassium move and change from one form to another from the soil profile into growing plant processes. Imbalances lead to the installation of deficiencies due to nutrient loss, affecting crop productivity. Cation exchange capacity is a measure of how much the soil can hold and a measure of the number of cations exchanged, such as calcium and potassium. High cation exchange capacity denotes good nutrient retention and slow release to plants, whereas low cation exchange capacity endows nutrients with an edge into leaching. Organic carbon should be taken into consideration, since it also plays a major role in the structure, water retention, and nutrient availability of the soil, which helps enhance fertility. The indicators portray, together, the soil health with its sustainable management practices.

#### *12.3.1.3 Biological Indicators*

Biological indicators are essential parts of soil health evaluation; they express the diversity and activity of life within the soil ecosystem, which drives several key functions for agriculture and environmental resistance. Apart from supporting improvement in the mechanisms of nutrient cycling, organic matter breakdown, and soil structure stabilization, a diversified microbial community will enhance resistance against all stressors and imbalances. Some of these include the soil enzymes dehydrogenase, urease, and phosphatase, acting as catalysts to underlie key processes, including nutrient cycling and organic matter breakdown. Increased activities of the enzymes show a vivid biological life in the soil. Soils of sufficient organic carbon quality can improve soil structure, allow for maximum moisture retention, increase nutrients available for plant growth, and are also one of the major carbon sinks to avert climate change. Earthworm populations are among the best indicators of soil health, as they can decompose organic matter, recycling nutrients, aerating the soil, and thus improving its structure. Such activities will form pore spaces in the soil, which in turn will favor root growth of many plant species. Beneficial bacteria, including phosphate solubilizers, increase a plant's chances of obtaining nutrients for growth while simultaneously inhibiting any pathogen. The high diversity of bacteria endows the soil ecosystem with its very existence to manage and its productivity. In unison, these biological indicators provide invaluable information about the state of soil health, guiding sustainable soil management.

### 12.3.2 Soil Sampling and Data Collection

The recognition of soil health monitoring is dependent on the accurate techniques of sampling and data collection. The objective is mainly to get representative samples of overall soil conditions from a field or region.

This is generally best accomplished with a combination of zone sampling, remote sensing, and soil probing for a comprehensive monitoring of soil health. Remote sensing provides broad insights, while zone sampling and probing offer targeted, depth-specific data. This hybrid approach balances accuracy, efficiency, and scalability. Zone sampling is based on the identification of distinct soil management zones; samples are taken from different areas with varying soil properties. Such practices stand valid within the context of precision agriculture since different zones may require specific interventions. Remote sensing asks for technologies like drones, satellites, and aerials in the provision of a fine resolution of data about soil properties. These data can be utilized in identifying patterns in soil health across extensive areas without the necessity for physical sampling. Soil probing and auger sampling means manual or powered soil probes are used to extract samples from specific depths, providing information on soil health at different levels, such as surface layers versus subsoil.

### 12.3.3 Data Normalization

There is no standardized way of normalizing agricultural data [17]. With agricultural data being so heterogeneous, taken from different sources that may have different units: soil moisture, temperature, and electromagnetic induction, one needs to normalize from any one point to some common scale; otherwise, working with them will not be easy. Normalization could solve the issue of integrating data by standardizing values across the spectrum. It will also ensure that some variables are not dominating the models due to their scale differences, and thus provide easy use for machine learning algorithms on the data. There are several methods of normalization available: min-max scaling, z-score standardization, and log transformation, depending on the nature of the data and analysis required. The best method is contingent on the specific nature of the agricultural data and modeling requirements. However, a commonly used and effective normalization technique for soil health data is min-max normalization.

This is a basic transformation method where each data value is rescaled into a specified range, generally [0, 1]. It entails giving each data value, after subtracting the corresponding minimum limit of the dataset, the quotient of the value range (max-min). Ideally, min-max normalization is where one needs to preserve relations between data points while involving soil health data (e.g., moisture levels, pH, or temperature) that have defined ranges that are uniform across all samples. This technique makes data comparable and can be easily interpreted.

$$X_{norm} = \frac{X - X^{min}}{X^{max} - X^{min}} \quad (12.1)$$

Where $X$ is the raw value, $X_{min}$ is the minimum value, and $X_{max}$ is the maximum value.

### 12.3.4 Integration of Sensor Data

The introduction of sensors into soil health monitoring has revolutionized soil measurement by providing real-time, continuous data across extensive farmland areas. This information is valuable to farmers for better decision-making to boost productivity and sustainability promptly. Handheld soil humidity sensors measure the moisture content in the soil at specific depths. This gives timely notice reports to make informed irrigation decisions and avoid inducing drought stress. Electromagnetic interference sensors offer a way of determining electrical conductivity to identify soil texture, salinity, and moisture. These assist in the mapping of soil variability across fields. Temperature sensors check the soil temperature, which affects seed germination, root development, and microbial activity, hence aiding in timely planting and management of soils. The pH and nutrients sensors gauge soil fertility in terms of nitrogen, phosphorus, and potassium, as well as pH, directly assisting in real-time fixing of fertilization methods. Near-infrared and mid-infrared radiation spectroscopic techniques permit fast and non-invasive determination of soil organic carbon and soil nutrient status, respectively, rapidly and accurately covering large areas. The IoT platforms contribute a way to connecting different sensors for off-site viewing and analysis. These platforms aim to combine soil data with weather forecast simulation platforms and farm management tools, providing farmers with a complete picture of field conditions. This integration permits decisions to be taken on data by spotting trends, predicting outputs, as well as through creating recommendations to improve soil management. With different sensor technologies combined with cloud computing and data analytics, soil health monitoring is more efficient and sustainable. This also allows farmers to optimize their inputs, decreasing their costs, and considering the long-term health of soils through sustainable agricultural production.

### 12.3.5 Reinforcement Learning

Reinforcement learning is a subset of machine learning. As depicted in Figure 12.2, an agent learns how to find solutions by collaborating with an environment and gains feedback in the form of appreciation or penalties. This RL agent takes actions within the environment to amplify cumulative rewards over time. In soil health surveillance, RL can be utilized to develop adaptive decision-making systems that improve soil management practices, optimize resource utilization, and enhance agricultural productivity. In RL, an agent networks with an environment to achieve the intended goal. The term of the agent in RL is based on doing a series of actions in a particular environment to obtain a reward. The key components of RL are:

- **Agent**: The decision-maker (in this case, a system for soil management) that takes actions to enhance soil health.
- **Environment**: An external system (the soil ecosystem) that the agent communicates with and consists of all factors, such as moisture, nutrient levels, temperature, and microbial activity, that potentially impact soil health.
- **State**: Representation of the current state of the environment. Such factors would have included information such as soil properties, moisture levels, nutrient content, and microbial activity in the final representation of soil health.
- **Action**: It is the type of decision-making or intervention that has been applied toward manipulating the environment. For soil health, this could refer to adding fertilizers, the timing of the irrigation application, or any other method, such as soil conservation techniques.

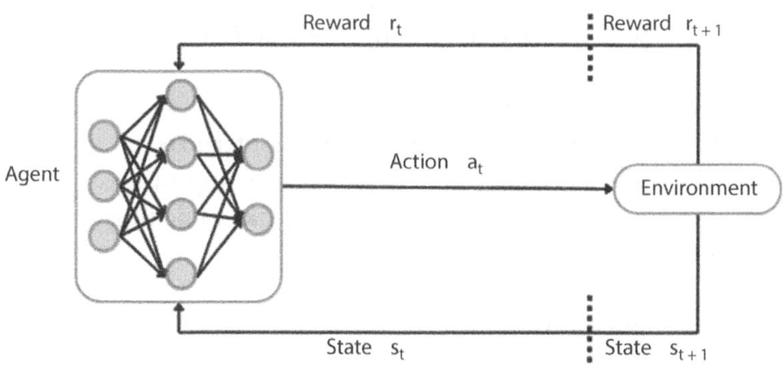


**Figure 12.2** Reinforcement learning.

- ❖ **Reward**: The feedback is given to the agent based on the actions it undertook. The reward reflects how well the action helped improve soil health, enhance productivity, or reduce resource usage.
- ❖ **Policy**: An approach or plan for states to take actions that the agent should follow to make choices. A good policy maximizes long-term rewards.

In RL, the agent iterates through several cycles of communication with the environment. Initially, it may take random actions, but over time, through trial and error, the agent gains knowledge about the best actions to take in diverse states to enlarge long-term rewards.

### *12.3.5.1 Designing the Learning Model for Soil Surveillance*

Designing an RL model for soil health surveillance involves structuring the problem so that the agent can continuously learn from its environment and improve its soil management decisions.

**❖ Defining the Environment**

Consider the environment as a soil system, disposing of itself dynamically under the influence of multifarious factors. For an RL model to represent the environment, real-time information gathered through soil sensors, such as moisture content, nutrient levels, and pH, must be collected and used as input. The data defines the state space multidimensional description of current soil conditions. For example, soil fertility, temperature, moisture levels, and microbial activity could be part of the state description.

**❖ Defining Actions**

The set of possible actions corresponds to the interventions or decisions that can be made to manage soil health. These actions might include:

- ❖ Adjusting irrigation schedules.
- ❖ Applying different types or quantities of fertilizers.
- ❖ Introducing organic amendments (e.g., compost).
- ❖ Implementing conservation practices like mulching or crop rotation.

**❖ Learning Algorithm**

Many various RL algorithms can be applied depending on the model complexity and goals. Among them, Q-learning, Proximal Policy Optimisation,

Deep Q-Networks (DQN), or Policy Gradient Methods are quite common. Through these algorithms, the agent learns an optimal policy by transmitting the actions made, receiving feedback, and then refining its decision-making strategy.

#### *12.3.5.2 Proximal Policy Optimization*

Proximal policy optimization (PPO) is an extensively used RL algorithm that offers a balance between computational productivity and stability in policy optimization. As represented in Figure 12.3, it is an on-policy, model-free RL method that plans to enhance the agent's policy iteratively to expand long-term rewards. The algorithm is grounded in the philosophy of policy gradient methods, which directly promotes the policy without needing an explicit model of the environment.

Proximal policy optimization addresses two significant challenges in RL: stability and efficiency. Alternative methods of RL faced challenges related to large, unstable policy updates or computational inefficiencies.

##### 12.3.5.2.1 Advantage Function

The advantage function $A(s_t, a_t)$ makes an indication of how much better the action $a_t$ taken in state $s_t$ is compared to the average action taken in that state. The advantage function plays an important role in guiding the optimization of the policy; that is, how it contributes to determining if an action is to be reinforced or penalized.

The most common estimation methods for the advantage function are generalized advantage estimation (GAE) or temporal difference errors. The advantage can be computed as:

$$A(s_t, a_t) = R_t - V_{(st)} \tag{12.2}$$

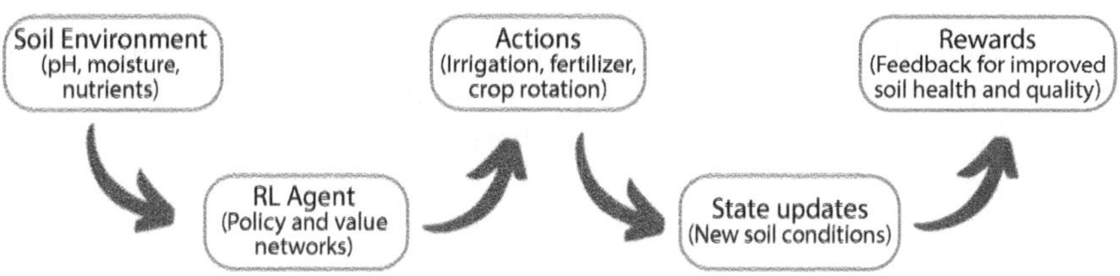


**Figure 12.3** Proximal policy optimization.

Where $R_t$ is the return (sum of discounted rewards) observed after taking action $a_t$, and $V_{(St)}$ is the predicted value of the state $s_t$ from the value function.

### 12.3.5.2.2 Surrogate Objective and Clipping Mechanism

In traditional policy gradient methods, the policy is updated by calculating the gradient of the expected return concerning the policy parameters. However, large updates to the policy can cause instability. The PPO mitigates this by restricting the size of the policy updates using the clipping mechanism.

The surrogate objective in PPO is defined as:

$$L^{clip}(\theta) = \Xi_t[min\,(r_t[\theta]A(s_t,a_t), clip(r_t(\theta), 1\text{-}\epsilon\ 1+\epsilon\ A(s_t,a_t)))] \tag{12.3}$$

here:

- $r_t(\theta)$ is the probability ratio between the new policy and the old policy:

$$r_t(\theta) = \frac{\pi_\theta(a_t \mid s_t)}{\pi_{\theta old(a_t|s_t)}}$$

- $A(s_t, a_t)$ is the advantage function, indicating how better the action $a_t$ is related to the baseline.
- The clip function restricts the probability ratio $r_t(\theta)$ to lie within the range $[1-\epsilon, 1+\epsilon]$, ensuring that the update is not too large.
- $\Xi_t$ is the expectation average of overall sampled transitions at time step *t*.

The clipped objective purpose ensures that the policy revision does not diverge meaningfully from the previous policy. This helps prevent the policy from changing too drastically, thereby maintaining stability in the learning process.

#### 12.3.5.2.3 Total Objective Function

The total objective function of the PPO combines several components to ensure efficient and stable learning. The ultimate objective is a weighted sum of the clipped policy loss, the value function loss, and an optional entropy term. The total objective function is given by:

$$L_{total}(\theta,\phi) = L^{clip}(\theta) + c_1 L_{VF}(\phi) - c_2 H(\pi_\theta) \tag{12.4}$$

Where:

- $L^{clip}(\theta)$ is the clipped surrogate objective.
- $L_{VF}(\phi)$ is the value function loss, which reduces the error between the forecast values and the observed returns.
- $H(\pi_\theta)$ is the entropy of the policy, which encourages exploration by penalizing deterministic behavior. The entropy term $H(\pi_\theta)$ is computed as:

$$H(\pi_\theta) = -\Xi_t[log\pi_\theta(a_t \mid s_t)]$$

- $c_1$ and $c_2$ are hyperparameters that regulate the comparative position of the value purpose loss and the entropy term.

The entropy term encourages the agent to maintain uncertainty in its actions, promoting the exploration of different strategies instead of prematurely converging to a suboptimal deterministic policy.

### *12.3.5.3 Algorithmic Workflow of the Proximal Policy Optimization*

The workflow of the PPO algorithm is defined in the following steps:

#### 12.3.5.3.1 Initialization

1. Activate the policy network $\pi_\theta(a_t \mid s_t)$ with random parameters θ.
2. Activate the value function $V_\phi(s_t)$ with random parameters ϕ.
3. Set the hyperparameters ϵ, $c_1$, $c_2$ and other relevant parameters.

#### 12.3.5.3.2 Data Collection

The agent cooperates with the environment by following the present policy to collect trajectories or episodes. During this phase:

1. The agent observes the current state $s_t$ of the environment.
2. Based on this $s_t$, the policy $\pi_\theta(a_t \mid s_t)$ is used to sample an action.
3. The agent performs the required action $a_t$ and transitions to the next state $s_{t+1}$, receiving a reward $r_t$.

This whole procedure is repetitive for a fixed number of time steps, during which the agent gathers experience.

#### 12.3.5.3.3 Advantage Estimation

ter collecting a batch of trajectories, the advantage function $A(s_t, a_t)$ is estimated for each action taken during the episodes. Typically, GAE is used to estimate the advantages. GAE reduces the variance of the advantage estimates while maintaining low bias.

#### 12.3.5.3.4 Policy Update

The PPO algorithm then updates the policy limitations θ by increasing the clipped objective function $L^{clip}(\theta)$. Simultaneously, the value function parameters ϕ are updated by minimize the value function loss $L_{VF}(\phi)$.

#### 12.3.5.3.5 Repeat

The process repeats by collecting new trajectories with the updated policy, estimating the advantages, and updating the policy and value function until convergence or the desired performance is achieved.

### *12.3.5.4 Proximal Policy Optimization for Soil Health Surveillance and Improved Product Quality*

Soil health surveillance involves the continuous monitoring of various soil limits such as pH, moisture content, organic matter, temperature values, and nutrient levels. By using PPO, agricultural systems may represent ongoing and adaptive modifications that can improve themselves continuously through the interplay of various dynamic factors affecting soil and the environment, to better manage soil health and product quality.

The PPO algorithm can be incorporated into any agricultural system, aiding decision-making pathways, arresting interventions for optimization purposes, and encouraging long-term soil health rehabilitation.

The PPO, an RL algorithm, is specially designed for this task, balancing exploration with exploitation while maintaining stability during learning. The PPO achieves balance through a clipped objective function, which prevents drastic changes to the policy during updates. This stability is implemented for applications, such as soil health management, where huge, sudden alterations in farming practices can have negative long-term consequences on both soil and product quality.

In the context of soil health surveillance, PPO is applied to continuously monitor and optimize the soil conditions. This process-based agent interacts with the environment by modifying farming activities based on the soil's initial condition and receives feedback in the form of different soil-health indicators, namely nutrient levels, pH, and moisture content. The agent will then learn the optimal set of interventions for improving soil quality and societal productivity and quality of agricultural products.

## 12.4 Conclusion

In this chapter, the importance of developing data-driven associations between soil health indicators and crop outputs is demonstrated. It is found that, upon the grant of RL, more informed decisions can be made to support moisture health, but also support crop yields and quality. The RL facilitates decision-making regarding sustainable agricultural practices on the principles of precision farming. Given the challenges associated with incorporating advanced soil health monitoring techniques, further exploration and incorporation of RL in agriculture have produced better results. Other issues like sensor accuracy, affordability, and accessibility related to the proposed methodologies can empower farmers, especially smallholder ones, to use the latest technology capabilities with confidence. Finally, soil health surveillance and RL integration provide new diversification spots, but the prospective benefits stretch beyond individual farms, supporting the global commitments of sustainability, resilience, and food security. As long as agrarian society continues to progress, incorporating innovative solutions such as RL will be key to delivering a healthier and more productive future for crops and the environment.

**AI disclosure statement:** The author confirms that no artificial intelligence (AI)-generated content or AI-assisted writing tools were used in the preparation of this chapter (including summaries and the abstract). All text, analysis, and conclusions are the result of the author's original work.

## References

1. Suresh Babu, C.V. and Yadavamuthiah, K., Soil quality prediction using deep learning, in: *Global Perspectives on Robotics and Autonomous Systems: Development and Applications*, pp. 171–188, IGI Global, 2023.
2. Wu, W., Li, Y., Yan, M., Yang, L., Lei, J., Liu, H.B., Surface soil metal elements variability is affected by environmental and soil properties. *PLoS One*, 16, 7, 2021.
3. Islam, M.R., Oliullah, K., Kabir, M.M., Alom, M., Mridha, M., Machine learning-enabled IoT system for soil nutrient monitoring and crop recommendation. *J. Agric. Food Res.*, 14, 100880, 2023.
4. Rajak, P., Ganguly, A., Adhikary, S., Bhattacharya, S., Internet of Things and smart sensors in agriculture: Scopes and challenges. *J. Agric. Food Res.*, 14, 100776, 2023.
5. Kurnaz, T.F., Dagdeviren, U., Yildiz, M., Ozkan, O., Prediction of compressibility parameters of the soils using artificial neural network. *SpringerPlus*, 5, 1, 1801, Springer Nature Switzerland, 2016.
6. Yogeshwari, M. and Prasanth, A., Smart Dust Technology for Monitoring and Control Systems in Smart Agriculture and Crop Surveillance Systems, in: *Computer Vision in Smart Agriculture and Crop Management*, pp. 123–137, Scrivener Publishing LLC and John Wiley & Sons, Inc., 2025.
7. Idowu, J., Pietrasiak, N., Hoellrich, M., *Soil biological processes*, New Mexico State University Cooperative Extension Service, Guide A-153, vol. 10, p. 8, 2023.
8. Ghimire, B., Spectroscopic sensor data fusion to improve the prediction of soil nutrient contents, Master's Thesis, Department of Biological Systems Engineering, University of Nebraska, Lincoln, 2023.
9. Biswas, A., Geostatistics in soil science: A comprehensive review on past, present, and future perspectives. *J. Indian Soc. Soil Sci.*, 72, 22, 2024.
10. Sarangi, A., Madramootoo, C.A., Cox, C., A decision support system for soil and water conservation measures on agricultural watersheds. *Land Degrad. Dev.*, 15, 49, 2004.
11. Kouadio, L. and Kone, M., Machine learning applications in agriculture: A comprehensive review. *J. Sens.*, 18, 2674, 2018.
12. Mohanty, M., Sinha, N.K., Painuli, D.K., Bandyopadhyay, K.K., Hati, K.M., Reddy, K.S., Chaudhary, R.S., Modeling soil water contents at field capacity

and permanent wilting point using artificial neural network for Indian soils. *Natl. Acad. Sci. Lett.*, 10, 171, 2015.
13. Lima, H.S., Oliveira, G.F.V.D., Ferreira, R.D.S., Castro, A.G.D., Silva, L.C.F., Ferreira, L.D.S., Oliveira, D.A.D.S., Silva, L.F.D., Kasuya, M.C.M., De Paula, S.O., Silva, C.C.D., Machine learning-based soil quality assessment for enhancing environmental monitoring in iron ore mining-impacted ecosystems. *J. Environ. Manage.*, 356, 120559, 2024.
14. Xu, X., Ma, F., Zhou, J., Du, C., Applying convolutional neural networks (CNN) for end-to-end soil analysis based on laser-induced breakdown spectroscopy (LIBS) with less spectral preprocessing. *Comput. Electron. Agric.*, 199, 107171, 2022.
15. Tripathi, A., Tiwari, R.K., Tiwari, S.P., A deep learning multi-layer perceptron and remote sensing approach for soil health-based crop yield estimation. *Int. J. Appl. Earth Obs. Geoinf.*, 113, 102959, 2022.
16. Pham, T., Acharya, P., Bachina, S., Osterloh, K., Nguyen, K., Deep-learning framework for optimal selection of soil sampling sites. *Comput. Electron. Agric.*, 217, 108650, 2024.
17. Abinaya, M. and Vadivu, G., Role of Big Data Analytics in Intelligent Agriculture, in: *Computational Intelligence in Internet of Agricultural Things*, pp. 19–48, Springer Nature Switzerland, 2024.

# 13

# Case Study on Automated Crop Disease Detection and Classification Using Computer Vision and Reinforcement Learning Techniques

**Fathima G.[1*], Sujatha S.[2], Raghu Ramamoorthy[3] and Pritha A.[4]**

[1]*Department of Computer Science and Engineering, Adhiyamaan College of Engineering, Hosur, Tamil Nadu, India*
[2]*Department of Electrical and Electronics Engineering, Sri Sairam College of Engineering, Anekal, Karnataka, India*
[3]*Department of Computer Science and Engineering, Oxford College of Engineering, Bengaluru, Karnataka, India*
[4]*Department of Science and Humanities, Anna University, Chennai, India*

### *Abstract*

Agriculture remains a cornerstone of the global economy, but plant diseases caused by bacterial, fungal, and environmental factors continue to threaten crop productivity and food security. Technology is becoming more crucial as it continues to transform contemporary farming into new methods for monitoring plant disease. This chapter suggests an online tool for precisely identifying plant diseases that makes use of convolutional neural networks, a subset of reinforcement learning. This system allows farmers to upload pictures of sick plant leaves. The technology detects plant diseases using segmentation and feature extraction by two contemporary image processing techniques. Strong performance and precise predictions are guaranteed by the system's extensive training on a range of plant leaf datasets. This device improves yield quality and quantity by enabling farmers to take necessary precautions against crop damage through early diagnosis. This chapter shows how reinforcement learning can revolutionize conventional agriculture by providing easily accessible, scalable, and effective disease management strategies.

**Corresponding author*: fathima.ace@gmail.com; ORCID: 0000-0002-3453-5359

Anitha Velu, Prasanth Aruchamy, Raghu Ramamoorthy, Rajesh Kumar Dhanaraj and Seifedine Kadry (eds.) Climate Smart Agriculture, (261–276) © 2026 Scrivener Publishing LLC

***Keywords:*** CNN, reinforcement learning, image preprocessing, VGG16 model, image augmentation

## 13.1 A Run-Through on Artificial Intelligence in Agriculture

### 13.1.1 Overview

As precision agriculture develops, the internet's exponential growth has made massive volumes of data available in various formats known as "big data," which can be analyzed and processed for a variety of purposes using reinforcement learning (RL) techniques. Reinforcement learning involves training neural networks using labelled datasets to learn from their surroundings and forecast without explicit programming. Reinforcement learning focuses on building models that can access data and learn to extract meaningful patterns and characteristics from it. New approaches are essential for effectively handling these challenges as the need for food production increases and plant disease poses a serious threat. Supporting a growing field of artificial intelligence (AI) in agriculture, which is a key driver of RL, was the aim of this study. Incorporating AI into agricultural practices aims to discuss a few main problems covering the comprehensive agricultural network while simultaneously promoting effective and sustainable crop cultivation methods.

### 13.1.2 Technological Advancements in Agriculture

Technological advancement in agriculture has significantly increased in its application. Modern farming practices, which are vital to human existence and civilization, have been adopted by knowledgeable farmers. The economic growth of nations often depends on agricultural trade. Among the various factors influencing agriculture, disease prediction is paramount due to current environmental conditions and human habits. Maintaining healthy crops amid changing environmental conditions is challenging, making disease prediction essential to prevent crop losses and starvation. Plant afflictions, leading to pathogenic organisms such as bacteria, fungi, nematodes, and viruses, hinder plant intensification and maturity, prior to diminishing the eminence and magnitude of plant products and premature plant death. Symptoms of plant diseases include nutrient deficiency (e.g., blossom end rot due to calcium deficiency), leaf spots caused by bacteria or fungi, discoloration (yellowing, mottling, chlorosis) caused by viruses, fertilizer burn, and pesticide burn.

## 13.2 Background of Artificial Intelligence in Climate-Smart Agriculture

Research has been carried out by focusing on the revealing of illness on cotton leaves using the K-means clustering algorithm [1]. The authors highlight that a significant proportion of cotton diseases manifest on the leaves, making early detection crucial. The methodology involves enhancing the contrast of input images through histogram equalization, followed by segmentation using the K-means clustering algorithm to classify the items based on explicit features. The definitive categorization is reached using a neural network. By making it smooth to label and understand leaf ailments early and accurately, this plan aims to assist laborers in directing and identifying potential crop deficits.

A hybrid support knowledge approach may be used to advance disease discovery in fruits and leaves [2]. This method aims to manage and identify plant ailments early and accurately. By joining CNN with an accompanying recurrent neural network (RNN), the authors upgrade the accuracy and productivity of disease. The pattern is evaluated using a dataset of figures of both healthy and damaged fruits and leaves, manifest by means of what it can promote a range of concerning farming by just and promptly recognizing defects. Nonetheless, by including advanced computational intellectual methods, many modern and contemporary technologies increase the exactitude of ailment labeling. By providing complete support for early disease discovery, this approach aims to reduce potential crop misfortunes and expedite proactive land practices. Image transform methods are essential for shooting plants to discover plant leaf ailments [4]. The approach involves preprocessing leaf images to embellish the contrast and engaging feature distillation methods to identify defects. Convolutional neural networks are more correct than established methods when it comes to categorization. To aid crop administration and prevent the possibility of yield losses, this method aims to manage plant disease early and correctly.

A novel policy that reinforces CSA is a domain-based CNN (R-CNN) model that makes and detects afflictions in vegetable plant leaves [5]. By labeling and classifying unhealthy domains in leaf photos, this plan admits the exact labeling of the number of ailments that impact potato plants. The R-CNN model's influence in early ailments labeling and discovery is shown by the difference in datasets used for preparation and judgment. This approach aims to correct agricultural practices, upholding direct crop administration and addressing potential yield misfortunes through precise and prompt plan affliction labeling.

Machine learning plans are important for detecting blast diseases in edible grain crops developed in South India [6]. The study stresses the significance of early detection in threatening crop misfortunes and lifting edible grain yield. The methodology used in the record of what happened includes shooting rice paddies and labeling appropriate countenance to change the middle from two points: unhealthy and acrobatic leaves. They analyze the dossier utilizing classification techniques such as K-nearest neighbors (KNN) and artificial neural networks (ANN). The studies show that an artificial neural network outperforms KNN and is more correct at changing middle from two points: healthy and sound leaves. This method aims to boost land productivity by guaranteeing convenient and correct disease identification.

Combining machine intelligence and representation to deal with forms has significantly revised the discovery of plant afflictions [7]. The study stresses the importance of early affliction diseases in threatening crop losses and lowering land productivity. Machine learning models are used to extract visual aspects and classify plant leaf figures; subsequently, they have been preprocessed. The methods used in this record of what happened explain the wonderful veracity in the expert's ability to improve crop administration practices and implement prompt interventions.

The most adept approach found might be the consolidation of concept separation and a multiclass support vector machine [8]. This approach accurately includes the range of ailments by utilizing a support vector machine to divide diseased extents in the vegetable leaf countenance. This arrangement proved active in recognizing ailments such as early and late blight, providing a trustworthy finish for early labeling and control. By reducing crop losses and permissive up-to-date interference, the system aims to boost land productivity. This inquiry boons a means for detecting plant petal affliction using color change forms. The research highlights the importance of early disease identification in preventing crop losses [9] and lifting land output. The technique includes preprocessing leaf concepts, extracting main features, and therefore utilizing a CNN to gain correct classification. This application provides farm and agricultural experts a constructive finish to embellish crop management plans and complete activity prompt invasions if necessary its extreme veracity in labeling a range of plant adversities.

To identify plant afflictions and evaluate their severity, researchers have been working on cosmopolitan figure preparation and RL methods [10]. This paper foregrounds a vital part of detecting high-fidelity disease is the progress of sustainable farming. The method contains a set of tomato early blight leaf images and CNN architectures, to a degree Alexnet, SqueezeNet, ResNet101, ResNet50, ResNet18, and Inception V3. Compared to duplicate disease discovery, this paper shows that ResNet101 acts as the best choice. Additionally, the images are classified into four asperity levels: Healthy, Early, Middle, and Late, accompanied by categorization accuracy ranging from 66.88% to 78.98%. This contemplation focuses on the potential of RL standards to boost agricultural productivity by accurately and quickly identifying plant diseases and determining their severity. The study's limitations [11] include the possibility of producing artificial images that may not accurately represent the range of real plant disease symptoms, which could affect model generalization. In practical scenarios, segmentation accuracy may be reduced by the method's [12] inability to handle complex backdrops or overlapping symptoms. One of the limitations of the research [13, 14] is that the deep CNN requires an enormous quantity of labeled records for training.

## 13.3 Detection and Classification of Plant Disease

To focus on the thriving need for more fastidious and effective defect disclosure patterns in farming, this case study aims to create cutting-edge RL

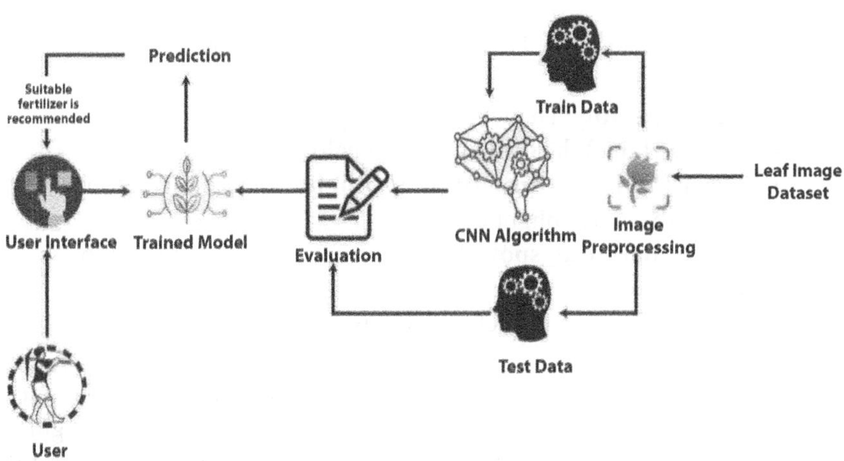


**Figure 13.1** System architecture.

systems for detecting plant leaf diseases. Furthermore, there are promising opportunities to address plant disease detection given RL's potential to transform visual interpretation and analysis. The intention is to develop a system that can meticulously recognize prevailing plant defects in leaves utilizing deep neural networks and illustration transform techniques, providing farmers with timely and applicable disease diagnosis information.

Starting from the leaf image dataset, the solution architecture in Figure 13.1 constructs two models: one for detecting diseases in the leaves of fruits like corn, peaches, apples, and another for detecting diseases in the leaves of vegetables like tomato, potato, and pepper plants.

### 13.3.1 Dataset Description

Two uncovering models, the crop dataset and the vegetable dataset, are the solution architecture proven in Figure 13.1, which establishes the leaf countenance of the dataset. These datasets, associated with this case study, determine inclusive image data that may be used to create healthy and reliable algorithms for diagnosing plant afflictions.

#### *13.3.1.1 Fruit Dataset*

The Fruit dataset shows the beginning of information in visible form, recycled within the study. It involves data classes, equally fruit.h5, which is a group of representations of various crop leaves recycled for test and training models. The following six types of fruit leaf images are included in this dataset:

- **Apple_Black_rot:** Depicts the apple leaves that are affected by black rot.
- **Apple_healthy:** Depicts healthy apple leaves.
- **Corn (Maize)_healthy:** Depicts healthy corn leaves.
- **Corn (Maize)_Northern_Leaf_Blight:** Depicts symptoms of northern leaf blight disease.
- **Peach_Bacterial_spot:** Depicts peach leaves affected by bacterial spots.
- **Peach_Healthy:** Depicts healthy peach leaves.

Inclusive of some healthful and unhealthy items in the selection expedites the comparison of the labeling process. For meticulously recognizing the disease affecting the plant and providing recommendations for appropriate disease control methods, such as fertilizer application, this comparison is essential.

### *13.3.1.2 Vegetable Dataset*

The Vegetable dataset is another serious fact used in this study. It has nine data classes, including images of various vegetable leaves. The design is trained and approved on these representations, which are referred to as vegetable.h5. Nine different types of vegetable leaf images are included in this dataset:

- Pepper_Bell_Bacterial_spot: Depicts bacterial spot infection.
- Pepper_Bell_Healthy: Depicts healthy pepper bell leaves.
- Potato_Early_blight: Depicts the symptom of early blight disease in the potato leaf.
- Potato_healthy: Depicts healthy potato leaves.
- Potato_Late_blight: Depicts the image affected by late blight disease.
- Tomato_Bell_Bacterial_spot: Depicts bacterial spot infection in the tomato leaf.
- Tomato_Late_blight: Depicts the symptom of late blight disease in tomato leaves.
- Tomato_Leaf_mold: Depicts the mold sign of the tomato leaf.
- Tomoto_Septoria_Leaf_spot: Depicts tomato leaves with septoria leaf spot.

Similar to the Fruit dataset, images of both active and unhealthy leaves are determined to aid in inaccurate disease detection. By correlating both representations, the model can effectively label the condition and advise appropriate situations based on distinguishing specific diseases affecting the vegetable plant.

A crucial, indiscriminate part of this case study explores increasing numbers of both active and afflicted leaves in the dataset. By training the model to recognize the performance that distinguishes healthy from diseased leaves. This technique embellishes the model's capability to induce and precisely identify diseases in everyday venues. By providing an inclusive group of representations, the datasets reveal the model to a wide range of disparities, enhancing its resilience and reliability.

The detailed depictions and distribution of the data underline the dataset's implication in processing plant disease detection models. The designed structure of the dataset improves training and testing efficiency and intensifies the overall potency and precision of the models in detecting plant leaf diseases.

### 13.3.2 Data Augmentation Techniques for Enhancing Model Generalizability in Reinforcement Learning

A key tactic in RL is data augmentation, which involves a few methods to produce "new" training samples from the original ones. These techniques preserve the class labels while introducing random oscillations and perturbations into the data. The primary model of data augmentation is to improve the model's generalizability, strengthening it and increasing its capacity to handle the variations of real-world data. The neural network can be exposed to slightly modified versions of the input data to learn more robust characteristics. Since data augmentation is not used during testing and the model is evaluated using the original testing data, higher testing accuracy is attained even though there may be a slight decrease in training accuracy.

#### *13.3.2.1 Random Jittering for Generalization*

A normal distribution with zero mean and unit variance can be used to demonstrate data augmentation. An accurate approximation of the distribution might result from training a machine learning model with this data. But in real-world applications, data rarely shows such a neat distribution. To increase the classifier's generalization, random oscillations are introduced to the data points by appending values chosen from a random distribution. Although the plot still follows an essentially normal distribution, it is not as flawless as the original one. The likelihood of a model generalizing to example data points not in the training set is higher when it is trained on this supplemented data. The model's efficacy in practical situations depends on this enhanced generalization.

##### 13.3.2.1.1 Types of Image Augmentation

In the realm of RL for computer vision, there are three primary types of data enlargement capabilities. The opportunity of approach depends on the definite background of the study or experiments.

#### 13.3.2.1.1.1 Type #1: Dataset Generation and Expansion (Less Common)

Dataset generation involves creating an expanded dataset from a single image by applying various transformations. This process includes:

a) Loading the original image from disk.
b) Applying random transformations such as translations and rotations to the image.
c) Saving the transformed image back to disk.
d) Repeating steps (b) and (c) multiple times.

This method results in a collection of transformed images based on a single input image. However, it is often more practical to gather additional data or explore behavioral cloning methods for dataset expansion.

#### 13.3.2.1.1.2 Type #2: In-Place/on-the-Fly Data Augmentation (Most Common)

The Keras ImageDataGenerator class is used to achieve this common type of data augmentation. This technique ensures that the network will encounter new data variations at each period. This method encompasses

i) Provide a batch of images to the ImageDataGenerator
ii) Rotate, translate, or otherwise change each image at random.
iii) Return modified batch to the calling function.

These modifications are made during training by the ImageDataGenerator, which sends the modified data back to the neural network without being aware of the network changes.

#### 13.3.2.1.1.3 Type #3: Combining Dataset Generation and In-Place Augmentation

This technique combines in-place augmentation with dataset reaction and is commonly applied in behavioral cloning. The ImageDataGenerator receives a batch of images, performs random image alteration, and outputs the updated data. In contrast to the Keras ImageDataGenerator, this method may also return the original batch along with the updated data. These techniques for data augmentation significantly enhance the generalizability and performance of RL models. These methods better handle the intricacy and unpredictability of real-world data, which would ultimately improve their accuracy and robustness in practical applications.

### 13.3.3 Employing Keras ImageDataGenerator for Image Preprocessing

The following steps outline the image preprocessing activities using the Keras ImageDataGenerator class, which was used in the implementation of the study:

i) Import the ImageDataGenerator Library.
ii) Configure the ImageDataGenerator Class.
iii) Pertain to the ImageDataGenerator functionality of the training and testing data sets.

The ImageDataGenerator generates the updated data after randomly altering the original data. A strong and diverse dataset is produced through the data augmentation process, which enhances the model's ability to generalize. Generic image processing techniques encompass a larger range of operations to improve image quality, extract features, or modify the image for specific purposes, in addition to data augmentation. The four main types of image processing techniques are as follows:

**Pixel Brightness Transformations/Brightness Corrections:** Entails changing an image's contrast or brightness to improve visibility and normalize lighting.
**Geometric Transformations:** Includes image standardization techniques, including cropping, translation, scaling, and rotation for additional analysis.
**Image Filtering and Segmentation:** Involves segmenting the image into meaningful parts to detach objects of interest and using filters to enhance or extract essential properties, such as edge detection and smoothing.
**Fourier Transform and Image Restoration:** Image restoration is done to repair the damaged image by deblurring and removing noise, whereas the Fourier transform is done to convert images from the spatial domain to the frequency domain for analysis. Data enlargement and comprehensive image transforming capabilities are imperative for embellishing the efficacy and generalizability of RL models. These tactics can be used to assemble more potent frameworks that scrupulously diagnose plant ailments, which will ultimately bear in more efficient and tenable breeding routines.

### 13.3.4 Feature Extraction Using a Pre-Trained VGG16 Model

Before training the CNN model, the dataset of potato leaf illustrations is contingent on feature extraction using the VGG16 pre-trained model [15]. The VGG16 model is well-designed for computer vision, surpassing in an extensive range of applications, subsequently being trained on the ImageNet dataset. The definitive classification layer was detached from the VGG16 model for this case study, leaving only the convolutional layers. These layers were then used to split features from the representations of potato plant leaves. The outcome of the ultimate convolutional layer was compressed and fed into the layers that were sufficiently associated. The performance of the customized CNN model is upgraded by leveraging the pre-trained VGG16 model's competence to derive intricate features from images. This method influences the power of many features that VGG16 has learned during its extensive training on the ImageNet dataset.

### 13.3.5 Model Training with a Convolutional Neural Network

In this study, a CNN is developed to classify plant leaves based on the type of disease. For the Fruit dataset, the categories included Apple Healthy, Apple Black Rot, Corn Healthy, Corn Northern Leaf Blight, Peach Healthy, and Peach Bacterial Spot. For the Vegetable dataset, the categories included Pepper Healthy, Pepper Bacterial Spot, Potato Early Blight, Potato Healthy, Potato Late Blight, Tomato Bacterial Spot, Tomato Late Blight, Tomato Leaf Mold, and Tomato Septoria Leaf Spot.

The CNN architecture consisted of four convolutional layers, followed by two fully connected layers, and ended with the final output layer having a SoftMax activation function to predict class probabilities [3]. The model comprised a total of 1,792 parameters. The Adam optimizer with a learning rate of 0.0001 and a batch size of 32 is used to train the model. For evaluation, the accuracy metric and categorical cross-entropy loss function were considered. The model's performance was monitored over the course of 50 epochs of GPU training using a validation dataset that made up 20% of the dataset. The CNN's affirmation accuracy of 95.8% signifies that it was profitable in classifying plant leaves into the relevant disease groups. This study builds engaging approaches for dropout regularization and data augmentation, which inevitably reduce overfitting and reinforce models' accomplishments. By following these steps, the pattern's dependability and ability to generalize to new, untested data were entrenched.

### 13.3.6 Model Deployment in Application

Following the profitable training and validation, the next vital step is to utilize the CNN model for real-world use. This deployment involved incorporating the trained model into a relevant system that used input images from the Fruit and Vegetable databases to distinguish and categorize plant diseases automatically. The demand had to handle a range of input image types and produce decisive and compatible results. The objective of this study was to create a comprehensible application that would assist farmers and other agricultural professionals in rapidly identifying and considering plant diseases. To ensure a smooth operation, the deployment phase considered the model's compatibility with different devices and environments. The goal of this integration was to turn the case study's findings into a practical solution for managing diseases in agriculture. By providing a quick and simple method for disease diagnosis, the model was meant to increase agricultural productivity and sustainability.

### 13.3.7 Implementation

The system's ability incorporates a web application, which is the front end for end-user interactions, a server at the backend for image processing and classification [5, 6], and a database for maintaining images and related information. Using a digital camera or similar equipment, different types of photographs are taken to identify the affected area of leaves. These images are then processed using a variety of image-processing techniques [10] to extract versatile and valuable information required for plant leaf disease analysis. Increasing image statistics is the goal of image preprocessing, which suppresses unwanted distortions and highlights image capabilities that are likely pertinent for similar processing. After receiving an input image, the preprocessing creates output images that are smoothed, inverted, and grayscale.

The Keras RL repository was used to carry the pre-trained VGG16 model and further extract features from the dataset. Additionally, Keras was employed to build and train the custom CNN model using the extracted features. Initially, the information preprocessing and feature abstraction steps were completed, and then it proceeded to the model training phase. The open-source platform, Jupyter Notebook, was utilized to construct the RL model, employing a CNN architecture due to the specific nature of the plant leaf disease identification task. With distributions of 80% and 20% respectively, the training, validation, and testing sets were first isolated from the dataset. Accuracy, the Adam optimizer, and the unconditional cross-entropy loss function were then employed as evaluation metrics. The training process was conducted with a batch size of 32 over 50 epochs using a CPU. At this stage, the accuracy and loss of the model on the training and justification sets were monitored using TensorBoard, a visualization tool catered by TensorFlow. To improve the model's performance, the factors like learning rate, batch size, and optimizer parameters were changed. When the model's routine was evaluated in the training set after training, its accuracy was 95.8%. The competent model weights were first stored in HDF5 format and then transformed to the TensorFlow Lite layout for implementation.

## 13.4 Results and Discussions

This segment offers a comprehensive evaluation of the RL-based plant leaf disease detection system. Extensive testing with some of the input images, as illustrated in Figure 13.2, and analysis demonstrate the CNN model's ability to identify and classify a range of plant diseases. This section elucidates the quantitative and qualitative aspects of the representation's routine in terms of the precision of disease detection. The integration of advanced feature extraction techniques with the custom-designed CNN architecture yielded exceptional results in disease classification. Specifically, transfer learning was implemented by leveraging pre-trained models, which significantly enhanced the model's feature extraction capabilities. This approach proved particularly effective in capturing subtle disease-specific patterns in potato plant leaf images, resulting in robust classification performance across different disease categories.

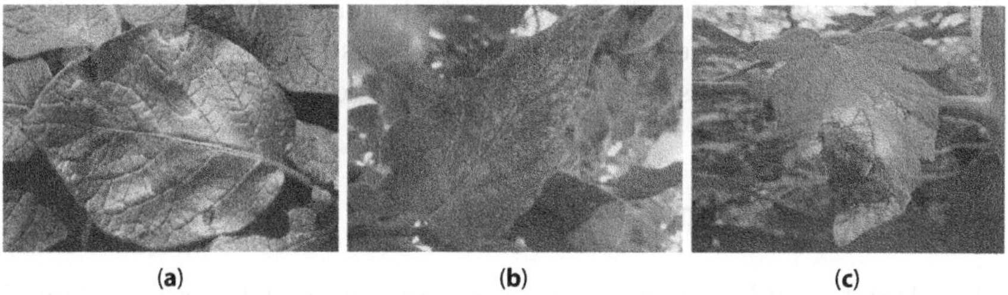

(a) (b) (c)

**Figure 13.2** Sample input leaf images with disease. (a) Potato early blight disease; (b) tomato spider mite disease; (c) tomato late blight disease.

```
Epoch 1/10
475/475 [==============================] - 2540s 5s/step - loss: 1.1399 - accuracy: 0.6122 - val_loss: 0.6337 - val_accuracy: 0.7750
Epoch 2/10
475/475 [==============================] - 187s 391ms/step - loss: 0.5642 - accuracy: 0.8006 - val_loss: 0.6368 - val_accuracy: 0.7777
Epoch 3/10
475/475 [==============================] - 166s 350ms/step - loss: 0.4399 - accuracy: 0.8416 - val_loss: 0.3436 - val_accuracy: 0.8781
Epoch 4/10
475/475 [==============================] - 173s 364ms/step - loss: 0.3510 - accuracy: 0.8739 - val_loss: 0.2538 - val_accuracy: 0.9099
Epoch 5/10
475/475 [==============================] - 171s 360ms/step - loss: 0.3088 - accuracy: 0.8903 - val_loss: 0.2739 - val_accuracy: 0.9050
Epoch 6/10
475/475 [==============================] - 172s 362ms/step - loss: 0.2832 - accuracy: 0.9004 - val_loss: 0.2343 - val_accuracy: 0.9201
Epoch 7/10
475/475 [==============================] - 168s 353ms/step - loss: 0.2486 - accuracy: 0.9127 - val_loss: 0.1633 - val_accuracy: 0.9435
Epoch 8/10
475/475 [==============================] - 184s 388ms/step - loss: 0.2266 - accuracy: 0.9205 - val_loss: 0.1554 - val_accuracy: 0.9465
Epoch 9/10
475/475 [==============================] - 178s 375ms/step - loss: 0.1995 - accuracy: 0.9300 - val_loss: 0.1892 - val_accuracy: 0.9345
Epoch 10/10
475/475 [==============================] - 179s 378ms/step - loss: 0.1949 - accuracy: 0.9309 - val_loss: 0.1207 - val_accuracy: 0.9580
```

**Figure 13.3** Result on the accuracy of potato plant disease detection.

The evaluation framework used in this study incorporated multiple performance metrics to ensure a comprehensive assessment. The primary metrics included confusion matrices for analyzing class-wise performance, precision and recall scores for measuring detection accuracy, and F1-scores for balanced evaluation. The model underwent rigorous testing on an independent test dataset, carefully curated to represent real-world scenarios and completely segregated from training and validation sets. It is observed in Figure 13.3 that a remarkable testing accuracy of 95.8%, validating the model's generalization capabilities and practical applicability.

Furthermore, this study conducted an extensive comparative analysis against contemporary state-of-the-art models in plant disease detection. These benchmarking experiments provided valuable insights into the model's competitive advantages and potential areas for enhancement. This study successfully applies dropout regularization and sophisticated data augmentation techniques, which is a significant contribution. In-depth ablation studies demonstrated how these techniques significantly decreased overfitting issues and enhanced the model's overall robustness.

## 13.5 Conclusion

The plant leaf affliction uncovering accomplishment favorably attained its basic objective, which was to cultivate a progressive, automatic disease labeling order utilizing RL principles. A meaningful innovation in agricultural technology, this contemporary system affords farmers with a decisive tool for meticulous and timely disease diagnosis, expediting enriched crop management practices and higher yield inflation. To sufficiently classify and evaluate the system's influence, a huge collection of plant leaf representations from diversified disease types and environmental circumstances was used. The overall accuracy was attained at 95.8%, with some disease categories reaching accuracy levels of up to 98%, manifesting the system's extraordinary competence to identify defects [3]. These remarkable results illustrate the system's reliability and feasibility in real-world agricultural scenarios. The benefit of the study was attributed to the strategic use of cutting-edge tools and technology throughout the progress lifecycle. As the pivotable component of the development schema, Jupyter Notebook improved the efficiency of data preprocessing, model training, and evaluation. Sophisticated Python Integrated Development Environments, exclusively Visual Studio and IntelliJ IDEA, were beneficial in the development and optimization of the code. Effective model serving was made possible by a dependable web server architecture that made use of FastAPI during the deployment stage. This carefully thought-out technological ecosystem ensured optimal system performance and deliverable quality. The study's recommendations manifest the promising potency of RL-based techniques for agricultural disease detection systems. The results apparently signify that these automated systems may be helpful tools for farmers, enabling them to implement better crop management practices and increase yields by identifying and addressing diseases early. Due to its high accuracy and stability, the system may be able to address contemporary agricultural challenges, especially those affecting crop and sustainable farming practices.

**AI disclosure statement:** The author confirms that no artificial intelligence (AI)-generated content or AI-assisted writing tools were used in the preparation of this chapter (including summaries and the abstract). All text, analysis, and conclusions are the result of the author's original work.

## References


1. Pawan, P.W. and Ganorkar, S.R., Detection of Diseases on Cotton Leaves Using K-Mean Clustering Method. *Int. J. Res. Eng. Technol.*, 2, 425, 2015.
2. Aishwarya, B.K., Moulana, M., Vasimunnisa, S., Shamshadh, U., Shareen, K., Noorullah, R.M., Enhancing Disease Detection in Fruits and Leaves Through a Hybrid Reinforcement Learning Approach, in: *Proceedings of 4th International Conference on Smart Electronics and Communication*, Institute of Electrical and Electronics Engineers (IEEE) pp. 898–906, 2023.

3. Rohan, T., Devendra, M., Ravi, Y., Disease Prediction in Plants: using new Era Technologies, in: *2023 International Conference on Computational Intelligence, Communication Technology and Networking*, pp. 63–66, 2023.
4. Kundu, R., Chauhan, U., Chauhan, S.P.S., Plant Leaf Disease Detection using Image Processing, in: *2022 2nd International Conference on Innovative Practices in Technology and Management*, pp. 393–396, 2022.
5. Minu, B., Raja, M., Vinoth, K.P., Suvitha, S., Latha, R., Potato Plant Leaf Disease Detection and Recognition Using R-CNN Model, in: *2023 3rd International Conference on Pervasive Computing and Social Networking*, pp. 605–609, 2023.
6. Ramesh, S. and Vydeki, D., Application of machine learning in the detection of blast disease in South Indian rice crops. *J. Phytol.*, 11, 31, 2029.
7. Kiran, A., Lokesh, N.S.K., Raj, M.S., Palvadi, S.K., Plant Disease Detection using Image Processing with Machine Learning, in: *2023 4th International Conference on Electronics and Sustainable Communication Systems*, pp. 1590–1595, 2023.
8. Islam, M., Dinh, A., Wahid, K., Bhowmik, P., Detection of Potato Diseases Using Image Segmentation and Multiclass Support Vector Machine, in: *2017 IEEE 30th Canadian Conference on Electrical and Computer Engineering*, pp. 1–4, 2017.
9. Sahana, M., Reshma, H., Pavithra, R., Kavya, B.S., Plant Leaf Disease Detection Using Image Processing, in: *Emerging Research in Computing, Information, Communication and Applications*, N.R. Shetty, L.M. Patnaik, H.C. Nagaraj, P.N. Hamsavath, N. Nalini (Eds.), pp. 103–125, Springer, Singapore, 2022.
10. Verma, S., Chug, A., Singh, A.P., *et al.*, Plant Disease Detection and Severity Assessment Using Image Processing and Reinforcement Learning Techniques. *SN Comput. Sci.*, 5, 83, 2024.
11. Nazki, S., Yoon, A., Fuentes, S., Park, D.S., Unsupervised image translation using adversarial networks for improved plant disease recognition. *Comput. Electron. Agric.*, 168, 105117, 2020.
12. Singh, V., Sunflower leaf diseases detection using image segmentation based on particle swarm optimization. *Meas.*, 140, 184, 2019.
13. Mishra, R., Sachan, C., Rajpal, D., Deep Convolutional Neural Network-based Detection System for Real-time Corn Plant Disease Recognition. *Sustain. Comput. Inform.*, 28, 100407, 2020.
14. Sharma, Y.P., Berwal, S., Ghai, W., Performance analysis of Reinforcement Learning CNN models for disease detection in plants using image segmentation. *Comput. Biol. Med.*, 113, 103397, 2019.
15. Simonyan, C. and Zisserman, A., Very deep convolutional networks for large-scale image recognition, in: *Proceedings of the International Conference on Machine Learning*, pp. 1–14, 2014.

# Index

# Also of Interest

## Check out these other related titles from Scrivener Publishing

*Sustainable Agriculture Production Using Blockchain Technology,* Edited by Rekh Ram Janghel, Rajesh Doriya, Jaykumar Lachure, and Yogesh Kumar Rathore, ISBN: 9781394248674. Revolutionize the agricultural supply chain with this essential guide, which provides the practical knowledge to leverage blockchain technology for transparency, traceability, and trust, alongside AI for overcoming modern farming challenges.

*Mathematical Modeling in Agriculture,* Edited by Sabyasachi Pramanik, Niranjanamurthy M., Ankur Gupta, and Ahmed J. Obaid, ISBN: 9781394233717. The main goal of the book is to explore the idea behind data modeling in smart agriculture using information and communication technologies and tools to make agricultural practices more functional, fruitful and profitable.

*Agroforestry,* Edited by Abhishek Raj, Manoj Kumar Jhariya, Arnab Banerjee, Ramesh Kumar Jha, and Krishan Pal Singh, ISBN: 9781394231133. This book offers comprehensive insights into the management of agroforestry for livelihood security and sustainable development in the tropics, addressing ecological interactions, productivity, and the monetization of carbon credits, while also outlining a future roadmap and policy challenges.

*Smart Agritech: Robotics, AI, and Internet of Things (IoT) in Agriculture,* Edited by Santosh Kumar Srivastava, Durgesh Srivastava, Korhan Cengiz, and Pramod Gaur, ISBN: 9781394302959. The main goal of *Smart Agritech: Robotics, AI, and the Internet of Things (IoT) in Agriculture* is to explore how emerging technologies such as robotics, artificial intelligence (AI), and IoT can be leveraged to improve efficiency, sustainability, and productivity in agriculture.

*DIGITAL AGRICULTURAL ECOSYSTEM: Revolutionary Advancements in Agriculture,* Edited by Kuldeep Singh and Prasanna Kolar, ISBN: 9781394242931. The book comprehensively explores the dynamic synergy between modern technology and agriculture, showcasing how advancements such as artificial intelligence, data analytics, and smart farming practices are reshaping the landscape to ensure food security in the era of climate change, as well as bridging the gap between cutting-edge research and practical implementation.

*THE DIGITAL AGRICULTURAL REVOLUTION: Innovations and Challenges in Agriculture through Technology Disruptions,* Edited by Roheet Bhatnagar, Nitin Kumar Tripathi, Nitu Bhatnagar and Chandan Kumar Panda, ISBN: 9781119823339. The book integrates computational intelligence, applied artificial intelligence, and modern agricultural practices and will appeal to scientists, agriculturists, and those in plant and crop science and management.

*HANDBOOK OF AGRICULTURAL BIOTECHNOLOGY: Volume I Nanopesticides,* Edited by Charles Oluwaseun Adetunji and Julius Kola Oloke, ISBN: 9781119836148. The book provides detailed information on the application of nanopesticides for the management of numerous agricultural pests and pathogens, as well as to foster an innovative pathway toward future sustainable agriculture and food systems.

*HANDBOOK OF AGRICULTURAL BIOTECHNOLOGY: Volume II Nanobioherbicides,* Edited by Charles Oluwaseun Adetunji, Julius Kola Oloke, ISBN: 9781119836155. The book provides a detailed examination of the application of nanobioherbicides that come from plants including information on the different metabolites derived from numerous plants that could become bioherbicides.

*HANDBOOK OF AGRICULTURAL BIOTECHNOLOGY: Volume III Nanofungicides,* Edited by Charles Oluwaseun Adetunji and Julius Kola Oloke, ISBN: 9781119836162. This book provides state-of-the-art information about recent advancements in the application of nanobiofungicides for effective management of post-harvest diseases and agricultural fungal diseases, including blights, mildews, molds, and rusts.

*HANDBOOK OF AGRICULTURAL BIOTECHNOLOGY: Volume IV Nanoinsecticides,* Edited by Charles Oluwaseun Adetunji and Julius Kola Oloke, ISBN: 9781119836179. The book provides detailed information about the application of repellent products that contain plant-based ingredients known as nanobioinsecticides, including the pesticide evaluation scheme guidelines for repellent testing, relevant information about the procedures to evaluate several repellent compounds, the development of new products that offer high repellency, and guidelines for consumer safety.

*ECORESTORATION FOR SUSTAINABILITY,* Edited by Arnab Banerjee, Manoj Kumar Jhariya, Surendra Singh Bargali, and Debnath Palit, ISBN: 9781119879718. With climate change and land degradation advancing at an alarming rate, the state of the global ecosystem and how it can be restored in our march toward sustainability is incredibly important, and this new, groundbreaking volume tackles these important issues head on.

*FUNCTIONAL POLYMERS IN FOOD SCIENCE, Part 1: Food Packaging,* Edited by Giuseppe Cirillo, Umile Gianfranco Spizzirri, and Francesca Iemma, ISBN: 9781118594896. Provides an extensive overview of the link between polymer properties and food quality. *NOW AVAILABLE!*

*FUNCTIONAL POLYMERS IN FOOD SCIENCE, Part 2: Food Processing,* Edited by Giuseppe Cirillo, Umile Gianfranco Spizzirri, and Francesca Iemma, ISBN: 9781118595183. Provides an extensive overview of the link between polymer properties and food quality. *NOW AVAILABLE!*

*ADVANCES IN FOOD SCIENCE AND TECHNOLOGY Volume 1,* Edited by Visakh, P.M., Sabu Thomas, Laura B. Iturriaga, and Pablo Daniel Ribotta, ISBN: 9781118121023. Written in a systematic and comprehensive manner the book reports recent advances in the development of food science and food technology areas. *NOW AVAILABLE!*

*ADVANCES IN FOOD SCIENCE AND NUTRITION Volume 2,* Edited by Visakh P.M., Laura B. Iturriaga, and Pablo Daniel Ribotta, ISBN: 9781118137093. This important book comprehensively reviews research on new developments in all areas of food chemistry/science and nutrition. *NOW AVAILABLE!*

*Applied Water Science Volume 1: Fundamentals and Applications*, Edited by Inamuddin, Mohd Imran Ahamed, Rajender Boddula and Tauseef Ahmad Rangreez, ISBN: 9781119724766. Edited by one of the most well-respected and prolific engineers in the world and his team, this is the first volume in a two-volume set that is the most thorough, up-to-date, and comprehensive volume on applied water science available today. *COMING IN SUMMER 2021*

*Sustainable Water Purification*, by Safiur Rahman and M. R. Islam, ISBN: 9781119650997. This is the only book that takes a zero-waste approach to propose 100% sustainable water purification techniques. *NOW AVAILABLE!*

## Books in the series, "Bioprocessing in Food Science"

*NUTRACEUTICALS FROM FRUIT AND VEGETABLE WASTE,* Edited by Vidisha Tomer, Navnidhi Chhikara, Ashwani Kumar, and Anil Panghal, ISBN: 9781119803508. Nutraceuticals from Fruit and Vegetable Waste, the latest volume in the series, "Bioprocessing in Food Science," is an essential reference on the valorization of fruit and vegetable waste and the state of the art for fruit and vegetable processing.

*NUTRITIONAL SCIENCE AND TECHNOLOGY: Concept to Application,* Edited by Tejpal Dhewa, Anil Kumar Puniya, and Anil Panghal, ISBN: 97811198089. This new volume in the groundbreaking series, "Bioprocessing in Food Science," Nutritional Science and Technology, is a comprehensive resource covering topics such as human nutrition, food safety concerns, and emerging technologies like nanotechnology, and it also addresses the need for innovative food products, contamination, and regulatory challenges in the global food market.

*HARVESTING FOOD FROM WEEDS,* Edited by Prerna Gupta, Navnidhi Chhikara, and Anil Panghal, ISBN: 9781119791973. An exciting new volume in the groundbreaking new series, "Bioprocessing in Food Science," Harvesting Food from Weeds focuses on the latest processes, industrial applications, and leading research on this important, evolving new area, for engineers, scientists, students, and other industry professionals.

*MICROBES IN THE FOOD INDUSTRY,* Edited by Navnidhi Chhikara, Anil Panghal, and Gaurav Chaudhary, ISBN: 9781119775584. This newest volume in the groundbreaking new series, "Bioprocessing in Food Science," focuses on the latest processes, industrial applications, and leading research on microbes in the food industry, for engineers, scientists, students, and other industry professionals.

*NOVEL TECHNOLOGIES IN FOOD SCIENCE,* Edited by Navnidhi Chhikara, Anil Panghal, and Gaurav Chaudhary, ISBN: 9781119775584. The third book in the series, "Bioprocessing in Food Science," covers the latest and greatest new technologies in food science.

*FUNCTIONAL FOODS,* Edited by Navnidhi Chhikara, Anil Panghal, and Gaurav Chaudhary, ISBN: 9781119775560. Presenting cutting-edge information on new and emerging food engineering processes, *Functional Foods*, the second volume in the groundbreaking new series, "Bioprocessing in Food Science," is an essential reference on the modeling, quality, safety, and technologies associated with food processing operations today.

*THERMAL FOOD ENGINEERING OPERATIONS,* Edited by Nitin Kumar, Anil Panghal, and M. K. Garg, ISBN: 9781119775591. Presenting cutting-edge information on new and emerging food engineering processes, Thermal Food Engineering Operations, the first volume in the new series, "Bioprocessing in Food Science," is an essential reference on the modeling, quality, safety, and technologies associated with food processing operations today.

Printed and bound by CPI Group (UK) Ltd, Croydon, CR0 4YY
17/09/2026
14964264-0002